Global Issues in Water Policy

Volume 32

Global Issues in Water Policy is now indexed in SCOPUS.

Policy work in the water sector has grown tremendously over the past two decades, following the Rio Declaration of 1992. The existing volume of water-related literature is becoming dominant in professional outlets, including books and journals. Because the field of water resources is interdisciplinary in nature, covering physical, economic, institutional, legal, environmental, social and political aspects, this diversification leads in many cases to partial treatment of the water issues, or incomplete analysis of the various issues at stake. Therefore, treating a whole host of a country's water resources issues in one set of pages will be a significant contribution to scholars, students, and other interested public. This book series is expected to address both the current practice of fragmented treatment of water policy analyses, and the need to have water policy being communicated to all interested parties in an integrated manner but in a non-technical language. The purpose of this book series is to make existing knowledge and experience in water policy accessible to a wider audience that has a strong stake and interest in water resources. The series will consist of books that address issues in water policy in specific countries, covering both the generic and specific issues within a common and pre-designed framework.

Caroline Lejars • Séverine Bouard
Editors

Water Policy in New Caledonia

Participative Co-building of Water Governance in a Decolonization Process

Springer

Editors
Caroline Lejars
CIRAD (French Agricultural Research Centre for International Development)
UMR GEAU (Univ Montpellier, CIRAD, INRAE, IRD, BRGM, AgroParisTech, Institut Agro)
Montpellier, France

Séverine Bouard
IAC (New Caledonian Institute of Agronomy)
Pouembout, New Caledonia

ISSN 2211-0631 ISSN 2211-0658 (electronic)
Global Issues in Water Policy
ISBN 978-3-031-88195-4 ISBN 978-3-031-88196-1 (eBook)
https://doi.org/10.1007/978-3-031-88196-1

This work was supported by CRESICA

This Springer imprint is published by the registered company Springer Nature Switzerland AG
The registered company address is: Gewerbestrasse 11, 6330 Cham, Switzerland

Acknowledgements

We deeply appreciate all contributors to this book for their solidarity and trust in this project, for their support and continuous engagement even during the violent events that devastated New Caledonia in 2024.

Thank you to Ariel Dinar and Stefano Farolfi for pushing this project and for remaining patient while waiting for the product.

Thanks to Nicola Thayil for reading the texts and for her support in the translation when needed, as well as acting as language reviewer.

Since the beginning of research in Social Sciences on Water management, with the Project Freshwater management in the Pacific led by Katherin Daniell and funded by the Fonds Pacifique and later, the Research project GOUTTE (Water governance on customary lands in New Caledonia) funded by CRESICA (Consortium for the research higher education and innovation in New Caledonia) many students have enriched the analysis and the participatory process presented in this book. We would like to acknowledge here all these people and thank them again for their valuable contribution: Kathleena Amo, Jérome Betrancourt, Adèle Bommier, Perrine Bouteloup, Léa Chevreux-Warode, Delphine Coulange, Cyril Goarant, Lucie Gosset, Oriane Grellier, Amar Imache, Hortense Lecercle, Charline Nekiriai, Laëtitia Nemoadjou, Marie-Eve Perrin, Olga Peytavi, Marine Pizette, Juliette Renaud, Jean-Emmanuel Rougier, Christophe Sand, Emmanuel Tjibaou.

For their love, patience and support, Caroline thanks Clémence and Maxime, and Séverine thanks Vincent, Armand and Diane.

Contents

Editors and Contributors

About the Editors

Caroline Lejars is a researcher at CIRAD and currently deputy director of a research unit on Water Management (UMR GEAU) in Montpellier. Her research focuses on the governance and management of hydrological territories mainly in North Africa and the Pacific. Most of her research is based on participatory approaches to support the organization of collective action. She has carried out and directed various research projects on groundwater governance, water contracts and arrangements and participatory planning for water policies. She was a lecturer at AgroParistech (2009–2010), associate professor at the National Agronomic Institute Hassan II in Morocco (2011–2016) and at the New Caledonian Institute of Agronomy (2016–2019).

Séverine Bouard is a Human Geographer (PhD). Her research focuses on assessing the extent of agriculture and hunting/fishing among Indigenous Pacific livelihoods in the context of the commodification of nature and emerging Indigenous discourses on nature and places. She focuses on the trajectories of people and territories, highlighting social change at work. She has taken part in, or led, more than ten research programmes on natural resource management and has developed a research practice deeply rooted in fieldwork. Her aim is to analyse and co-design public policies with stakeholders that are based on and respectful of indigenous ontologies and livelihood strategies. Together with Caroline Lejars, they led the research programme GOUTTE (Water on Governance on Customary Lands in New Caledonia), on which this edited volume is based. She will join the Department of Environmental Management at Lincoln University (NZ) in May 2025.

Contributors

Aude Arrighi DAVAR, Directorate of Veterinary Food and Rural Affairs, Water Division, Nouméa, New Caledonia

Mahée Autunno IRD (French National Research Institute for Development), UMR Espacedev (IRD, Univ Montpellier, Univ Guyane, Univ La Réunion, Univ Antilles, Univ Nouvelle Calédonie, Univ Perpignan), Centre IRD, BP 05, Nouméa, New Caledonia

Sylvain Barone INRAE (Research Institute for Agriculture, Food and Environment), UMR G-EAU, Univ Montpellier, AgroParisTech, CIRAD, Institut Agro, IRD, Montpellier, France

Marion Bois DAF, Planning and Land Management Directorate, Province Nord, Kone, New Caledonia

Adèle Bommier CIRAD (Agricultural Research Center For International Development), UMR GEAU, Univ Montpellier, IRD, CIRAD, INRAE, BRGM, Institut Agro, AgroParisTech, Montpellier, France

DGTM, General Directorate of Territories and the Sea of Guyane, Cayenne, Guyane

Séverine Bouard IAC (New Caledonian Institute of Agronomy), Pouembout, New Caledonia

LU, Lincoln University, Lincoln, New Zealand

Victor David IRD (French National Research Institute for Development), Institut Méditerranéen de Biodiversité et d'Écologie (IMBE Marseille), Marseille, France

Anne-Laure Dotte UNC, University of New Caledonia, Equipe de Recherche Eralo, BP R4, Nouméa, New Caledonia

Julien Drouin IAC (New Caledonian Institute of Agronomy), Equipe SOLVEG (SOL et VEGétation), Station de Recherche Agronomique de Atha, BP 1407 La Roche, Mare, New Caledonia

Charlotte Duval DASS, Directorate of Sanitary and Social Affairs, Government of New Caledonia, Nouméa, New Caledonia

Gérard Fallon DAVAR, Directorate of Veterinary Food and Rural Affairs, Water Division, Nouméa, New Caledonia

Nils Ferrand INRAE, Research Institute for Agriculture, Food and Environment, UMR GEAU, Montpellier, France

Lucie Gosset IRD (French National Research Institute for Development), UMR Espacedev (IRD, Univ Montpellier, Univ Guyane, Univ La Réunion, Univ Antilles, Univ Nouvelle Calédonie, Univ Perpignan), Centre IRD, BP 05, Nouméa, New Caledonia

Angélline Goue Government of New Caledonia, DIMENC, New Caledonian Geological Survey (SGNC), Nouméa, New Caledonia

Sonia Grochain IAC (New Caledonian Institute of Agronomy), Nouméa, New Caledonia

Pierre-Yves Le Meur IRD (French National Research Institute for Development), UMR SENS (IRD, CIRAD, UPVM), Centre IRD, BP 05, Nouméa, New Caledonia

Caroline Lejars CIRAD (French Agricultural Research Centre for International Development), UMR GEAU (Univ Montpellier, CIRAD, INRAE, IRD, BRGM, AgroParisTech, Institut Agro), Montpellier, France

Vincent Mardhel BRGM (Geological Research Mining Bureau), Nouméa, New Caledonia

Pierre-Louis Mayaux CIRAD (Agricultural Research Centre for International Development), UMR G-EAU, Univ Montpellier, AgroParisTech, CIRAD, INRAE, Institut Agro, IRD, Montpellier, France

Olivier Monge Government of New Caledonia, DIMENC, New Caledonian Geological Survey (SGNC), Nouméa, New Caledonia

Charline Nekiriai IAC, New Caledonian Institute of Agronomy, Nouméa, New Caledonia

Marie-Eve Perrin IRD (French National Research Institute for Development)-University of West Brittany (UBO), University Paul Valéry Montpellier (UPVM), Nouméa, New Caledonia

Olga Peytavi IRD (French National Research Institute for Development)-IAC (New Caledonian Agronomy Institute), CIRAD (Agricultural Research Centre for International Development), University Paul Valéry Montpellier (UPVM), Centre de Recherche Nord-Thierry Mennesson, Pouembout, New Caledonia

IAC, CIRAD, IRD, Centre de Recherche Nord-Thierry Mennesson, Pouembout, New Caledonia

Léa Russ Government of New Caledonia, DIMENC, New Caledonian Geological Survey (SGNC), Nouméa, New Caledonia

Catherine Sabinot IRD (Research Institute for Development), UMR Espacedev (IRD, Univ Montpellier, Univ Guyane, Univ La Réunion, Univ Antilles, Univ Nouvelle Calédonie, Univ Perpignan), Centre IRD, BP 05, Nouméa, New Caledonia

Karine Saumë DASS, Directorate of Sanitary and Social Affairs, Government of New Caledonia, Nouméa, New Caledonia

Emmanuel Tjibaou Culture Department, Province Nord, Kone, New Caledonia

Jacques Wamejonengo IAC (New Caledonian Institute of Agronomy), Equipe SOLVEG (SOL et VEGétation), Station de Recherche Agronomique de Atha, BP 1407 La Roche, Mare, New Caledonia

Pearl Winchester DAVAR, Directorate of Veterinary Food and Rural Affairs, Water Division, Nouméa, New Caledonia

Geoffroy Wotling DAVAR, Directorate of Veterinary Food and Rural Affairs, Water Division, Nouméa, New Caledonia

Abbreviations

ADCK	Agency for the Development of Kanak Culture
ADRAF	Land Development and Planning Agency
AEP	Drinking water supply
AFD	French development agency
AFM	Association Française des Maires/French Association of Mayors
ANU	Australian National University
BDLISA-NC	Aquifer Systems Limits Database of New Caledonia
BDSS-NC	Subsoil Database of New Caledonia
BRGM	Geological Research Mining Bureau
CES	Economic and Social Council (CES) of New Caledonia
CESE	Environment, Economic and Social Council of New Caledonia
CIRAD	French Agricultural Research Centre for International Development
CNRT	National Centre for Technological Research
COOPLAGE	Coupling open and participatory tools to allow stakeholders to adapt them to water management
COOPLAN	Co-construction and Co-planning of an action plan for water management
CRESICA	Consortium for the Research, Higher Education and Innovation in New Caledonia
DAF	Directorate of Land and Planning
DAFE	Directorate of Agriculture, Forest and Environment
DASS	Directorate of Sanitary and Social Affairs
DAVAR	Directorate of Veterinary Food and Rural Affairs
DGTM Guyane	General Directorate of Territories and the sea of Guyane
DIMENC	Directorate of Industry, Mines and Energy in New-Caledonia
DPF	Public river domain
EDF	European Development Fund

ENSO	El Niño-Southern Oscillation
FLNKS	Kanak and Socialist National Liberation Front
GDP	Growth domestic product
GDPL	Special local law groups (Groupement du droit particulier local)
IAC	New Caledonian Institute of Agronomy
ICPE	Classified Installations for Environmental Protection (French Environment Code)
INRAE	French National Research Institute for Agriculture, Food and Environment
IRD	French National Research Institute for Development
IRSTEA	National Research Institute of Science and Technology for Agriculture and Environment
ISEE	Institute of Statistics and Economic Studies
IWRM	Integrated Water Resources Management
MEAE	French Ministry of Europe and Foreign Affairs
METEXPO	Name of a research project on impregnation levels and determinants of human exposure to metals
MISE	Interdepartmental Water Mission
MOOC	Massive Open Online Course
OECD	Organisation for Economic Co-operation and Development
OEIL	Environmental observatory
OFB	French Biodiversity Agency
PEP	Shared Water Policy/Politique de l'eau partagée
PGSSE	Water Safety Management Plan
PIF	Pacific Islands Forum
PROTEGE	Oceania regional project of territories for sustainable management of ecosystems
PUD	Urban Planning Plan
RPCR	Rally for Caledonia in the Republic
SAGE (Island Province)	Water management support system
SAGE	Water development and management service
SANC	Archives of New Caledonia
SEEM	Maré water and environment company
SGN	New Caledonia Geology Service
SIVOM	Multi-purpose intercommunal association
SPC	Pacific Community
STENC	Energy transition plan of New Caledonia
TCA	Short-term agricultural tax/Taxe conjoncturelle agricole
TSPA	Tax in support of agricultural production/Taxe de soutien aux productions agricoles
UNC	University of New Caledonia
UNDRIP	United Nations Declaration on the Rights of Indigenous Peoples

UNPFII	United Nations Permanent Forum on Indigenous Issues
VKP	Voh-Koné-Pouembout (3 municipalities of the Northern Province)
WHO	World Health Organization
WMC	Water Management Committee
WPP or WPA	Water Protection Perimeters/Area
WSP	Water Safety Plan
WWF	World Wildlife Fund

Chapter 1
Introduction, Objectives, and Scope

Séverine Bouard and Caroline Lejars

Abstract This chapter provides the overall context and setting for the volume. Firstly, a general overview of water issues and policy in New Caledonia is provided. Then the scope of the book is established along with the way each chapter addresses its objectives. It also provides an overview of the chapters, underlying the links and continuity among them.

Keywords New Caledonia · Water policy · Customary water governance

1.1 Overview of Water Issues and Policy in New Caledonia

1.1.1 Geography of New Caledonia & Population

New Caledonia is located in the South Pacific region, nearly 17,000 km from France, East of Australia (1400 km from Brisbane) and North of New Zealand (1800 km from Auckland).

This archipelago is composed of the Grande Terre (500 km long and 50 km wide) with the Loyalty Islands (Ouvéa, Lifou, Tiga, Beautemps-Beaupré and Maré) to the East, the Isle of Pines to the South and the Belep archipelago to the North. With a surface area of 18,575.5 km^2 and an exclusive economic zone of more than 2.4 million km^2, 74% of the population lives in the Southern Province, particularly

S. Bouard (✉)
IAC (New Caledonian Institute of Agronomy), Pouembout, New Caledonia

Lincoln University, Lincoln, New Zealand
e-mail: severine.bouard@iac.nc

C. Lejars
CIRAD (French Agricultural Research Centre for International Development), UMR GEAU (Univ Montpellier, CIRAD, INRAE, IRD, BRGM, AgroParisTech, Institut Agro), Montpellier, France
e-mail: caroline.lejars@cirad.fr

C. Lejars, S. Bouard (eds.), *Water Policy in New Caledonia*, Global Issues in Water Policy 32, https://doi.org/10.1007/978-3-031-88196-1_1

in the Greater Nouméa area, which accounts for 66% of the total population, compared with 19% in the Northern Province and 7% in the Island Province.

The updated 2019 census records 271,407 inhabitants of New Caledonia (269,000 inhabitants according to the 2014 census). A particularity of New Caledonia is that its population has very different cultural identities. According to the 2014 census, 39% of the population declare themselves to be "Kanak", i.e., belonging to the indigenous people who have been present on New Caledonian soil for 4000 years; 27% are "European" or "Caldoches" and the remaining 34% do not belong to either of these two groups and are originally from Polynesia, Asia, are of metis or declare themselves to be "Caledonian" (19,000 people in 2014).

New Caledonia's climate is subtropical and there is an unequal distribution of rainfall along an East-West axis due to the presence of the central mountain range on Grande Terre: the windward coast and the Central Range (East) are wetter and the vegetation more luxuriant, while the leeward coast (West) is drier and the vegetation is more of the niauli savannah type. Although water is a relatively abundant resource in the territory in relation to the population, there are strong issues to be addressed due to its unequal distribution in space and time (Schéma d'Orientation de la Politique de l'Eau Partagée NC, 2019): the problem of supplying quality drinking water to all inhabitants, particularly in rural areas, which are very sparsely populated and difficult to access (more than 7 out of 10 households do not have access to water and out of the 413 catchments, 31% produce untreated water and 33% have a vulnerable catchment area); preserving the quality of the aquatic environment, both on land and in the sea, particularly due to erosion, river embankment (600 km of rivers are over-engraved) and river pollution (waste, mines, agriculture); the use of water to ensure and develop the island's food sovereignty and the control of water and related risks in a context of urban and demographic development (19,000 buildings are in flood-prone areas).

Pressure on water is growing with regard to the distribution of water uses between the different sectors, even if the situation in New Caledonia cannot be described as conflictual at present.

1.1.2 Few Figures on Economic and Environmental Issues

The New Caledonian economy has experienced a remarkable increase in its overall wealth over the period 1965–2010: more than 4% annually, which is more than the growth of metropolitan France. On average, GDP per capita grew by 2.3% per year over the period 1990–2010; this growth has slowed down significantly since the mid-2010s, the slowdown in nickel prices (from 2014) and then the referendums and the health crisis. GDP per capita is now close to that of mainland France, ten to fifteen times higher than that of the Pacific islands closest to New Caledonia (Fiji, Samoa, Vanuatu), and twice that of the French overseas departments.

Pressure on water is increasing with regard to the distribution of water uses between the different sectors, even if the situation in New Caledonia cannot be described as conflictual at present. Depending on the geographical area of New Caledonia, water uses are distributed differently according to the activity sector: freshwater supply, industrial needs (nickel mining activities) and agriculture. The use of water for agriculture represents 70 million m^3/year against more than 80 million for drinking water supply.

The provinces and the government planned to extend the irrigated agricultural surfaces in order to improve the agricultural autonomy of New Caledonia. For instance, in the Northern province there is an important multi-use dam project that aims to meet the domestic and environmental water needs of the Voh-Koné-Pouembout (VKP) area and to ensure the irrigation of 4300 ha of crops. Furthermore, Water for industry and mining represents, according to the New Caledonian water department (DAVAR), 150,000 m^3 withdrawn per day. Mining concessions are divided between the North and South Provinces (51% in the North and 49% in the South) and are New Caledonia's main economic resources for export, with mining accounting for 95% of exports in 2017. However, 24% of the Water Protection Perimeters (WPP) overlap with mining concessions, which leads to tensions between industrial activity and securing water quality (Fallon et al., 2019). Mining interests will therefore have to be taken into account in the water governance system without compromising strategic drinking water resources and the quality of catchment areas.

1.1.3 Institutional and Political Setting

New Caledonia has been an overseas collectivity with a special status since the constitutional revision of 20 July 1998, which followed the signing of the Nouméa Agreement of May 1998. The status of New Caledonia in relation to the French Republic has evolved since 1853, when Napoleon III took possession of the archipelago on behalf of France, to the present day. Until 1946 New Caledonia had the status of a French colony where the code de l'indigénat was applied from 1887.

With this code, the rights of the indigenous peoples were greatly reduced (dispossession of land, forced labour, specific taxes, restriction of movement). With the end of the code de l'indigénat in 1946–1947, French nationality was granted to all French people, including the natives, but universal suffrage was not introduced until 1957 by decree. During this period, nickel mining expanded and led to waves of population migration, particularly from Vietnam, Japan, Indonesia and Wallis, and then with the nickel boom in 1969–1972 a policy of European settlement was implemented. From the 1960s to 1990, New Caledonia became increasingly autonomous through a succession of laws specifying the transfer of powers to the councils or assemblies of New Caledonia.

At the same time, conflicts between independentists and non-independents became more and more acute and violent. The debate on New Caledonia's independence from France was formalised with the creation of the Front Indépendantiste in 1979, which has been called the Front de Libération Nationale Kanak et Socialiste

(FLNKS) since 1984. The laws of 1984 ("Lemoine" statute), 1985 ("Fabius-Pisani" statute) and 1986 ("Pons" statute) followed one another in a climate of very strong tensions, period called "The events". Finally, the Matignon-Oudinot Agreements were signed between the French State and the two majority parties in New-Caledonia (FLNKS and the Rassemblement pour la Calédonie dans la République, RPCR). This agreement formalised the creation of the three current provinces (North, South and Loyalty Islands) and gave them their own powers, particularly in planning, development and environmental matters. While a referendum was planned for 1989, following the various agreements and in response to the tensions in the territory, a negotiation process between independentists, non-independentists and the State was initiated with the Nouméa Agreement of 5 May 1998, based on the establishment of a strong autonomy for New Caledonia.

The negotiation includes the organization of a maximum of three successive referendums concerning the access to full sovereignty of New Caledonia.

The Nouméa Agreement prefigures the elements that are enshrined in the organic law n°99-209 of 19 March 1999, notably on customary civil status and the recognition of the Kanak as the indigenous people of New Caledonia, the transfer of State competences to the institutions of New Caledonia as well as the political autonomy of its institutions (possibility of adopting country laws, election of provincial assemblies for example, creation of the Customary Senate).

These developments are accompanied by a certain economic prosperity given the high nickel prices from the 2000s. To this day, New Caledonian industry is based on mining activities and remains very sensitive to variations in mining prices and policies.

New Caledonia is divided into three provinces with a directly elected assembly and their own competences, such as adopting their own environmental code. In addition to the Congress, composed of elected representatives of the provincial assemblies, New Caledonia has the following institutions: the government, which has an executive role; the Customary Senate, composed of 16 senators from the 8 customary areas, and the Economic, Social and Environmental Council, which have advisory roles. There are 33 communes whose boundaries coincide with those of the customary areas.

1.1.4 Fragmentation of Water Responsibilities and Complex Water Governance: The Current Institutional Framework for Water

This division of responsibilities and consequently of roles and responsibilities between central and local institutions is complex and was still recently the subject of a political debate between the government and the North and South provinces regarding the delegation of water resource management that has been effective since 1997 (Table 1.1).

Table 1.1 Distribution of water competences in New Caledonia. Source: DAVAR (2018)

French state	Public health: if mayor fails to act				
New Caledonian Government	Public law and public river domain	Public hygiene and health (DASS department)	Water protection perimeters: DUP, defining needed prescriptions	Civil security (to New Caledonia since 2014) and natural risk management	Non-delegated water management: crossing structures, dams Since 2021, management of watercourses: abstractions, maintenance and protection of watercourses and riversides
Provinces (*3)	Environment: regulation of freshwater fishing (Southern Province), water abstraction (Northern Province), ICPE authorization (facilities requiring environmental impact assessment)		Regulation of sanitation		
Municipalities (*33)	Public health: pollution, flooding, breach of dykes		Drinking water supply and sanitation: distribution, setting up of water protection areas		
Customary Senate and Customary Councils (8 customary areas)	Given the exclusion of customary lands from the public domain, it is the customary authorities that are responsible for managing watercourses.				

The North and South Provinces have sent back the delegated competence since 2021, because no funding was associated with this delegation. Concerning the division of roles and competences with the customary people, this is still unclear and the implementation of the revision of the organic law of 1999 (following the Nouméa Accord of 1998) and particularly Article 44 is not complete.

This article states that 'The domain of New Caledonia includes, in particular and except on customary land: property that is vacant and without a master, including securities, shares and cash deposits that have become statute-barred within the time limits laid down for the State, those of persons who die without an heir or whose estates have been abandoned. It also includes, subject to the rights of third parties and except on customary land, watercourses, lakes, groundwater and springs. New Caledonia (government) has jurisdiction over its own territory or that of its provinces (Article 22–31 organic law n°99-209 1999) and jurisdiction over water resource management was delegated to the North and South provinces by deliberation 238/CP of 18 November 1997.

The management of water resources on customary land is the responsibility of the customary authorities but the provinces, after taking customary acts, can still operate on these lands and ensure the management of the resource with, for example, light maintenance operations or, in conjunction with the government and the commune, the establishment of a water protection perimeter for water catchment with a view to making it drinkable. The intervention of the Northern and Southern Provinces on customary land is in practice carried out after obtaining agreements with the customary authorities in the form of a customary act. However, this practice is not governed by up-to-date regulatory texts or financial agreements between the government, the provinces and the customary authorities. Particularly in the area of water management, arrangements between these different levels of administration are in practice favored at the local level, but also undermined by current budgetary cutbacks.

1.1.5 The Place of Customary Lands and Indigenous Rights in this Fragmented Water Governance

Customary land, as defined in the organic law of 1999, corresponds to almost 30% of total land. Spatially, customary land corresponds to only part of the customary areas. There is a clear difference between the delimitation of strictly customary areas and areas of customary influence, where Kanak claims to the land are still heard and where prior application before intervention can be made (even if the planned operation is on public land).

The land issue and the distribution of land between public, private and customary status is currently a source of debate, particularly on the issue of the water-land continuum and the restitution of land to the customary people displaced following the colonization of the territory of Grande Terre in 1853. The Land Development and Planning Agency (ADRAF) has the task of implementing this policy of returning land to the tribes following their claim to the land link.

In fact, water management on these lands raises questions about the financial capacity, human resources and regulatory frameworks available to the Kanak for water resource management on their territories, but not only. Management on customary land also raises questions about the possibility of unifying management and governance throughout the territory of New Caledonia without distinction between land tenure systems in line with the principle set out in the Water Shared Policy Plan for "A single water statute, independent of land considerations, guaranteeing the general interest and applicable nationwide".

1.2 Objectives and Scopes of the Book

At the beginning of 2018, only a few months before the first referendum on full sovereignty, the Government of New Caledonia launched the construction of the country's first water policy. Only a few months later, after a broad consultation that involved about 1 out of 600 New Caledonians, the government adopted a detailed policy strategy and an action-planning document with nearly 700 actions.

This participatory process has been supported and led by two official government representatives in charge of sectors directly connected to emerging water issues in New-Caledonia: Agriculture and Customary Affairs. This political situation is original because the two government representatives come from two strongly opposed parties. The government representative of Agriculture is part of the non-independantist movement whereas the Kanak government representative of Customary Affairs is part of the pro-independence movement. Despite this ideological gap, they succeed in actively supporting the process and making it to the end. This policy is now called the Shared Water Policy, the first version of the country law relative to the water public domain and to water protection is currently under public consultation.

Meanwhile, Caledonian research institutes, mainly through the Cresica (Consortium for the research higher education and innovation in New Caledonia) and the CNRT (National Centre for Technological Research) Nickel and Environment, have launched large research programs that focus on water. Funded by the New Caledonian Government, local collectivities and mining companies (concerning the CNRT), these programs aim at supporting water governance, water management on customary land, the evolution of the regulatory corpus in a context of cross competences water-environment-agriculture-health, the definition of standards (reserved flow, discharges, etc.) but also the search for solutions adapted to the financing of public policies.

The writing of this book is motivated by the aspiration to share the results of these research programs as well as the lessons learned from the water policy making process. This book will therefore be of interest to two different profiles of readers. Firstly, those interested in research on 'indigenous peoples' rights to water will find an original, and so far, unaddressed in New Caledonia, illustration of the interconnections between indigenous peoples' political, land and environmental claims and claims to ownership of the resource (Hidalgo et al., 2017; Boelens & Seeman, 2014; Gupta et al., 2013).

The consequences associated with the differences in institutional regimes for access to quality drinking water described by Penn (2016) will be outlined, as well as the innovative ways in which these differences are being addressed in New Caledonia. For researchers and experts working in regions where the symbolic weight of freshwater for island communities' matters, this book describing the New Caledonian case will complement the historical and anthropological analyses already available on the Pacific (Wagner & Jacka, 2018). It will show how the sacred and symbolic dimension of water is concretely translated into innovative arrangements for managing the resource, as in New Zealand (Jackson & Altman, 2009; Hsiao, 2012; Salmond, 2014; David, 2017). Secondly, this book will be of interest to researchers working on the development of participatory methodologies for more inclusive water policy and governance processes, particularly in cross-cultural and potentially conflictual contexts. Though stakeholders and the public are increasingly solicited by governments and public bodies to participate in the public decision process, the public is rarely involved in policy design (Colorni & Tsoukiàs, 2018; Ferretti et al., 2019). More and more literature and practice recommend involving the public right from the early stages of the policy cycle, i.e., issue identification, definition of the policy objectives and policy design (Barbier, 2005). This book thus gives an account of a large-scale participatory process in a multicultural situation, allowing the meeting of different stakeholders, including citizens, in the management of water and territories. It describes the process and the tools that supported the construction of the policy, and how existing participative frameworks (Hassenforder & Ferrand, 2021) have been adapted to fit the needs of this policy design. It shows how a large-scale participatory process can lead to action proposals that fit people's priorities and that are innovative and collective. Territorial managers (technicians, engineers, project managers, union directors, etc.) who implement participatory approaches in the framework of projects, plans or programs may therefore be interested in this book.

1.3 Navigating the Book

The book is divided into 4 parts and 15 chapters. Figure 1.1 shows the logic behind the book structure and organization.

The chapters can be read independently, each reveals and explains the linked results of researches and studies, each has its own message, arrived at through research on a thematic coherence. Grouping them into parts is the second level of structural coherence, of the development of arguments linking various disciplines.

Chapters 2, 3, 4, and 5 sets the overall context and includes a description of water resources and flows, an overview of water quality issues, a presentation of the balance between resources and uses, and a discussion of current legal framework related to water (part 1).

Chapters 6, 7, 8, and 9 assess and try to understand why, despite the relative abundance of the resource, tensions over water actually exist? In those chapters, the authors consider water management as a co-production of nature and societies (Aubriot & Riaux, 2013; Budds & Linton 2014) and we will use the notion of hydro-social water cycle to recognize, formalize and examine how cultural and historical dimensions of kanak water management are integrated in or arranged with modern water management. Such an approach will lead us to ask the question of the commensurability (incommensurability) of water forms (Li, 2014), but also the interactions between power relations and infrastructural dimensions in water control (Obertreis et al., 2016).

Mobilizing this conceptual framework in the Pacific, where indigenous peoples have developed a particular conception of Human-Nature relations is of additional heuristic interest. Each chapter proposes an analysis of how past modes of management and social organization around water meet contemporary public policies and governance of water. In practice, authors demonstrate that arrangements exist but without modern managers fully perceiving the underlying local values and

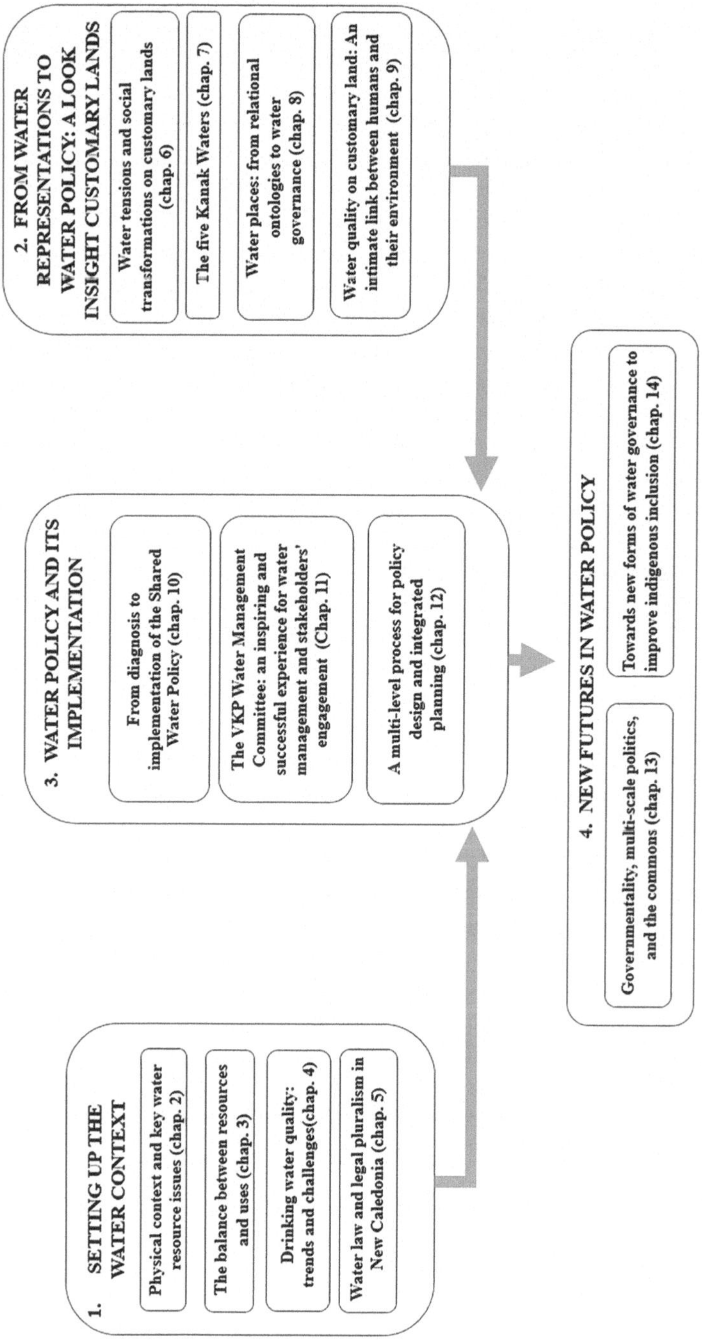

Fig. 1.1 Structure of the book

representations. It is clear that when water management is relatively efficient and without tension, it is because indigenous values and representations are partly respected, even if they are almost never made explicit or fully understood (part 2).

Chapters 10, 11, and 12 aimed at sharing the wide participatory experience led in New Caledonia. The three chapters present a retrospective of the consultation mechanism set up for the development of this policy. It focuses in particular on the proceedings of the process and on the planning, support tools that were used, partly inspired by the co-planning mechanism, called COOPLAN (Co-construction and Co-planning of an action plan for water management). Three different and complementary points of view are proposed: in chapter 10, the government officials in charge of water management look back at the consultation process launched in 2017, its ambitions, the results but also the limits and difficulties they encounter despite the participatory success. In Chap. 11, officials from the North Province return to the origins of the process at the local level, around the Pouembout catchment area and the growing issues of sharing the resource between water supply and agriculture, especially in view of the construction of a dam. Indeed, between 2015 and 2017, some New Caledonian water service agents were trained in the COOPLAN approach for the management committee of the VKP zone (Voh Kone Pouembout, in northern New Caledonia) (Pizette 2020; Bouteloup, 2016). With the support of researchers, some water service agents, previously trained in the approach or involved in the VKP water management committee, have thus remobilized these tools as part of the construction of the PEP. Finally, in Chap. 12, researchers adopt a more critical posture assessing and focusing on the process and on the planning, support tools that were used. Authors finally analyze the feedback from participants and facilitators and discuss the limitations of the tools (part 3).

Chapters 13, 14 and 15 question and explore the Future of Water Policy in New Caledonia and wrapped-up the main lessons that can be learnt from this case study in the Pacific. The Chap. 13 offers an opportunity to examine the process from a reflexive and critical point of view, and to discuss different kinds of management. The Chap. 14 exposes and discusses several governance models that could improve the acknowledgment and recognition of indigenousness, as a basis for the construction of a coherent water governance model on institutional, financial, operational and human levels. We finally conclude that the case of New Caledonia could be a source of inspiration for policy makers especially in countries where endogenous and exogenous populations and/or legal orders coexist. The last chapter offers a conclusion with the key-messages of the different chapters and proposes ideas for improving water governance in countries facing similar issues of indigenous claims and revendications for taking part to water governance.

References

Aubriot, O., & Riaux, J. (2013). Savoirs sur l'eau: les techniques à l'appui des relations de pouvoir? *Autrepart, 65*, 3–26.

Barbier, R. (2005). Quand le public prend ses distances avec la participation. *Natures Sciences et Sociétés, 13*(3), 258–265. https://doi.org/10.1051/nss:2005040
Boelens, R., & Seemann, M. (2014). Forced engagements: Water security and local rights formalization in Yanque, Colca Valley, Peru. *Human Organization, 73*(1), 1–12. https://doi.org/10.17730/humo.73.1.d44776822845k515
Bouteloup, P. (2016). La gouvernance de l'eau à partir d'une démarche participa-tive: de l'analyse à la planification de la gestion de l'eau sur la zone VKP, Nouvelle Calédonie, Dissertation IAC/PN, in collaboration with UMR G-eau, ISTOM.
Budds, J., & Linton, J. (2014). The hydrosocial cycle: Defining and mobilizing a relational-dialectical approach to water. *Geoforum, 57*, 170–180.
Colorni A., Tsoukiàs A. (2018) What is a decision problem? Designing alternatives, in N. Matsatsinis, E. Grigoroudis, Preference disaggregation in multiple criteria decision analysis, : Springer, p. 1-15.
David, V. (2017). La Nouvelle Vague des Droits de la Nature. La Personnalité Juridique Reconnue Aux Fleuves Whanganui, Gange et Yamuna. *Revue juridique de l'environnement, 42*(3), 409–424. https://droit.cairn.info/revue-juridique-de-l-environnement-2017-3-page-409?lang=fr
Fallon, G., Wotling, G., Chalagiraud, B., & Winchester, P. (2019). *Schéma d'orientation pour une Politique de l'Eau Partagée de la Nouvelle-Calédonie* (p. 196).
Ferretti, V., Pluchinotta, I., & Tsoukiàs, A. (2019). Studying the generation of alternatives in public policy making processes. *European Journal of Operational Research, 73*(1), 353–363. https://doi.org/10.1016/j.ejor.2018.07.054
Gupta, J., Pahl-Wostl, C., & Zondervan, R. (2013). 'Glocal' water governance: A multi-level challenge in the anthropocene. *Current Opinion in Environmental Sustainability, 5*(6), 573–580. https://doi.org/10.1016/j.cosust.2013.09.003
Hassenforder, E., & Ferrand, N. (2021). Évaluer une démarche participative. *Sciences Eaux & Territoires, 35*, 90–95. https://doi.org/10.14758/set-revue.2021.1.14
Hidalgo, J. P., Boelens, R., & Vos, J. (2017). De-colonizing water. Dispossession, water insecurity, and Indigenous claims for resources, authority and territory. *Water History, 9*(1), 67–85. https://doi.org/10.1007/s12685-016-0186-6
Hsiao, E. C. (2012). Whanganui river agreement: Indigenous rights and rights of nature. *Environmental Policy and Law, 42*(6), 371–375.
Jackson, S., & Altman, J. (2009). Indigenous rights and water policy: Perspectives from Tropical Northern Australia. *Austl. Indigenous L. Rev., 13*(27).
Li, T. M. (2014). What is land? Assembling a resource for global investment. *Transactions of the Institute of British Geographers, 39*, 589–602.
Obertreis, J., Moss, T., Mollinga, P., & Bichsel, C. (2016a). Water, infrastructure and political rule: Introduction to the special issue. *Water Alternatives, 9*(2), 168–181.
Penn, H. J. (2016). *Water security in the rural North: Responding to change, engineering perspectives, and community focused solutions*. PhD Dissertation University of Alaska Fairbanks.
Salmond, A. (2014). Tears of Rangi: Water, power, and people in New Zealand. *HAU: Journal of Ethnographic Theory, 4*(3), 285–309. https://doi.org/10.14318/hau4.3.017
Pizette, M. (2020). *L'évaluation des processus de concertation pour la construction des politiques de l'eau en Nouvelle-Calédonie* (p. 91) [Mémoire de stage pour l'obtention du Master 2 Gestion Territoriale et Développement Local]. Université Jean Moulin Lyon 3.
Wagner, J. R., & Jacka, J. K. (2018). *Island rivers: Fresh water and place in Oceania*. ACT ANU Press.

Part I
Setting Up the Context

Chapter 2
Physical Context and Key Water Resource Issues in New Caledonia

Olivier Monge, Léa Russ, Vincent Mardhel, and Angélline Goue

Abstract Despite an average annual accumulation of 1850 mm, the high variability of rainfall, the distribution of the population and activities but also the numerous constraints subject New Caledonia to recurring imbalances between resources and uses. Surface water is naturally more vulnerable, although its use is largely predominant in New Caledonia. They often appear critical, while the qualitative and quantitative challenges of groundwater make them strategic resources with potential that remains largely to be clarified and developed. This chapter describes and analyzes the numerous existing constraints and presents issues concerning water resources.

Keywords Surface and groundwater resources · Peridotites hydrosystem

2.1 Physical Context of Water Resource in New Caledonia

2.1.1 Geology

The onshore geology of New Caledonia cannot be separated from its regional and offshore context, from the eastern Australian coast to the Tonga–Kermadec Trench. This mostly submerged system of basins, ridges and seamounts of various ages is considered to be made up of oceanic and thinned continental crust, continental ridges, volcanic arcs, intra-plate volcanic chains and oceanic plateaus. Part of this region, including New Caledonia, is a submerged continent referred to as Zealandia (Mortimer et al. 2017).

O. Monge (✉) · L. Russ · A. Goue
Government of New Caledonia, DIMENC, New Caledonian Geological Survey (SGNC), Nouméa, New Caledonia
e-mail: olivier.monge@gouv.nc; angelline.goue@gouv.nc

V. Mardhel
BRGM (Geological Research Mining Bureau), Nouméa, New Caledonia
e-mail: v.mardhel@brgm.fr

C. Lejars, S. Bouard (eds.), *Water Policy in New Caledonia*, Global Issues in Water Policy 32, https://doi.org/10.1007/978-3-031-88196-1_2

New Caledonia consists of several islands that are the emergent parts of two parallel submarine ridges: the Norfolk Ridge, including Grande Terre to the south-west, and the Loyalty Ridge to the north-east.

Grande Terre is a complex mosaic of Phanerozoic geology (Maurizot et al., 2020a), including ophiolitic units, monotonous volcaniclastic sedimentary rocks, and volcanic and metamorphic terranes. The oldest rocks are Carboniferous. The most distinctive and emblematic geological unit of Grande Terre is the 5400 km^2 Peridotite Nappe, which is one of the largest ultramafic terranes on Earth. Weathering of the peridotites in the Late Cenozoic resulted in the formation of a thick lateritic regolith with significant supergene nickel deposits, as well as impacts in terms of water resources and natural and health risks (trace metals).

The three main Loyalty Islands are the emergent part of the Loyalty Ridge. They are largely composed of uplifted thick Miocene to Present carbonate platform deposits. Minor basalts are exposed on the island of Maré.

The Matthew and Hunter islands are volcanoes located on the southern tip of the active Vanuatu island arc, whereas the Chesterfield islets are atolls built on subsided intra-oceanic volcanoes of the Lord Howe seamount chain.

2.1.2 *Orohydrography*

Between longitudes 163°30′E and 168°15′E and latitudes 19°30′S and 22°45′S, New Caledonia has three main ensembles (Bonvallot, 2012): Grande Terre and its mountainous relief, the low islands of the Loyalty archipelago, and the two islands prolonging Grande Terre, the Belep group in the north and the Isle of Pines in the south.

Grande Terre extends over 400 km from south-east to north-west, and around 50 km in width, bordered by a 24,000 km^2 lagoon. The relief is formed by a central ridge with mountain peaks rising over 1000 m, divided into two high ranges: Humboldt (peaking at 1618 m) to the south-east and Panié (reaching 1629 m) to the north-west.

The mountains reach the sea very steeply, and watercourses are short. They cross small alluvial plains before flowing into the sea in branching estuaries, often in deep bays surrounded by steep slopes.

The western landscape of Grande Terre there is marked by coastal plains separated from each other by clearly marked hilly ranges with eroded steep slopes. Towards the sea these plains become small deltas occupied by mangroves. The coast is lower than on the eastern side, where mangrove fringes are also often found. Sometimes there are small, raised islands scattered across the lagoon, as in Saint Vincent Bay.

Isolated mountains situated in the piedmont area dominate the plains and the hill areas. They are often over 1000 m in altitude, lower towards the north.

The Belep islands to the north and the Isle of Pines to the south are prolongations of the central ridge.

The Loyalty Islands are formerly raised carbonate platforms, sometimes tilted by tectonic activity. The islands are flat with a depression in the centre. The altitude reaches some 100 m at most, and Ouvea's peak is only 45 m. They are bordered by coral limestone cliffs ranging from 20 to 50 m in height. These cliffs show shelves and notches from past sea levels, as well as many caves.

The watercourses on Grande Terre, which are nearly all oriented perpendicular to the general line of the island, end first of all in narrow alluvial plains, and then estuaries on the east coast, and small deltas on the west coast. The 55 km long Diahot River is the longest river in New Caledonia.

The Loyalty Islands have no surface waterways, with water circulation being exclusively underground due to the permeable coral limestone that make up their structure.

The modeled hydrographic network (MHN), developed from a digital elevation model, was established by the DAVAR[1] water service and distributed online (https://georep.nc/). An unofficial alternative hydrographic reference framework (MHN1ha) and associated analysis tools have been proposed (Goué et al., 2022) with the aim of helping to determine strategic surface resources. It includes a network schedule, and a nomenclature structured as follows: watershed, river, section, shore and body of surface water.

2.1.3 Weather, Cyclones and Climatic Features

New Caledonia is situated in the subtropical zone and its climate, characterised as tropical oceanic varying with the seasons and the phases of the Southern Oscillation (ENSO). This conditions the frequency of occurrence of different types of weather and the associated meteorological conditions (Maitrepierre, 2012).

Certain weather types are fairly constant through the year, for instance the trade winds, while others are more seasonal, such as the tropical regime and the westerly current.

The climatic year can be divided into two main seasons and two intermediate seasons:

- The warm season from November to April, often hot, damp and rainy.
- The cool season, from May to October. Lower rainfall and temperatures can be fairly low (under 10 °C). Cold fronts from the south can reach New Caledonia. August to October is also the driest season in the year.
- The transitional seasons between the two main seasons do not lend themselves to clear characterisation.

[1] Gouvernement de la Nouvelle-Calédonie, Direction des Affaires Vétérinaires Alimentaires et Rurales, Nouméa 98800, Nouvelle-Calédonie.

In New Caledonia climatic variability depends on several factors, the main ones being:

- Geographical factors, which influence all parameters and in particular precipitation and temperature. The dominant winds (eastsouth-east trade winds) and the relief on Grande Terre result in rainfall being much higher on the windward coast (east) and in the central mountain range than on the west coast. Mean temperatures decrease with altitude, while proximity to the sea moderates extreme values and the temperature range. The synoptic winds are also markedly affected by the local environment;
- The observed path of the sun;
- The climatic phases of ENSO (El Niño-Southern Oscillation) which influence interannual variations are estimated to account for around 30%. The El Niño episodes result in drier and cooler periods, while the La Niña episodes are warmer and wetter than normal.

The interannual average precipitation is around 1680 mm (Romieux, 2011), but with geographical variations range from 800 mm per year on the western coastline to more than 3500 mm annually on the eastern coastal peaks (Fig. 2.1). In terms of seasonality, the north of Grande Terre and the Chaîne record almost 50% of the

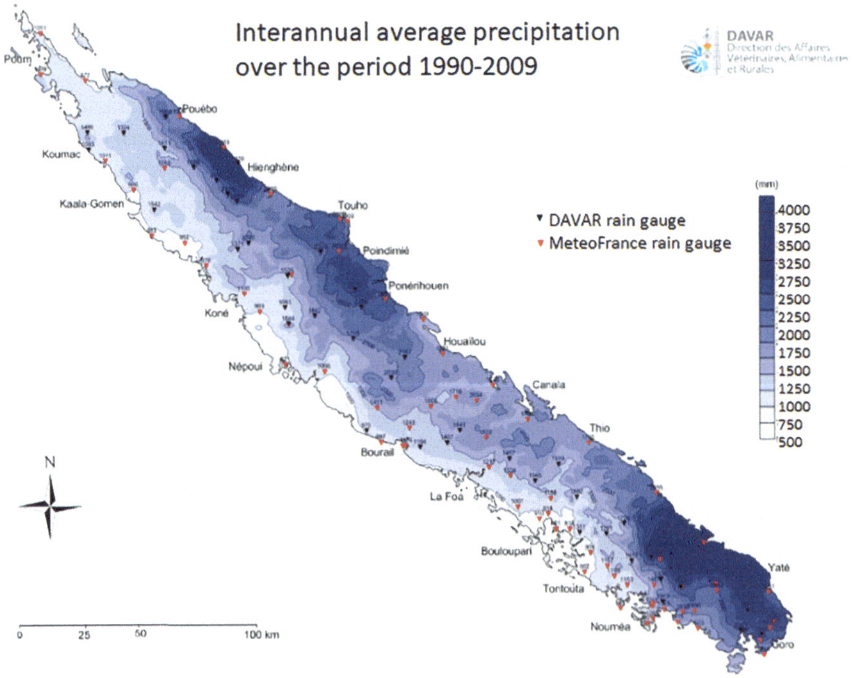

Fig. 2.1 A map of interannual isohyets in Grande-Terre (according to Romieux, 2011)

year's precipitation between the months of January and March. The seasonal contrast is less marked in the south of Grande Terre and the Loyalty Islands.

Although rare, tropical depressions and cyclones (Fig 2.2), formed in the warm season above the tropical Pacific, are a major climatic risk in the country. They carry significant precipitation, both in terms of volume and intensity. Accumulations >500 mm in 24 h are frequently recorded (913 mm/24 h for Cyclone Anne in 1988, Wotling, 2012). The torrential rain and high winds that accompany them cause considerable damage along their paths. Indeed, the high intensities or accumulations of rain associated with these phenomena result in remarkable floodings and are also the main triggering factor for slope instabilities. Especially since the peridotite massifs are the site of destructive torrential lavas, landslides (houses, roads, catchments) or deadly ones (8 victims on 11/22/2016 in Houaïlou). A landslide hazard mapping program is set to be completed in 2024 and to date has made it possible to record more than 37,500 phenomena in 15 municipalities with an average density of 3.6 phenomena per km^2.

Another phenomenon taken into account is: potential evapotranspiration that arises from high rainfall variability, extremely low water levels or extraordinary rainfall events. Evapotranspiration follows a very clear seasonal cycle between the months of the cool season, where values are the lowest, and those of the dry season where they reach their maximum (Maitrepierre, 2012). It also varies depending on

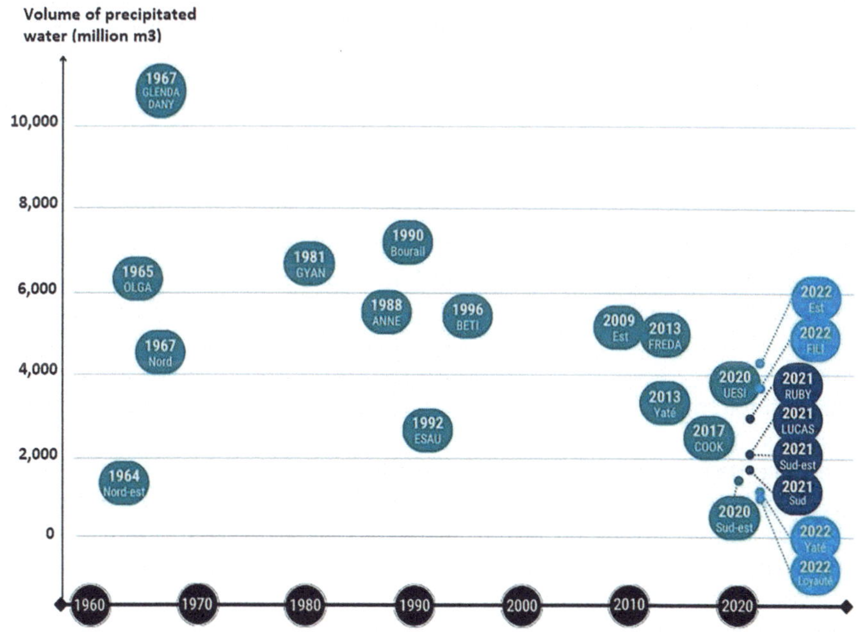

Fig. 2.2 Remarkable rainfall events in New Caledonia (1961–2022) (http://pluiesextremes.meteo.fr/nouvelle-caledonie/-Evenements-memorables-.html)

local conditions. The highest values are recorded on the west coast with a maximum in Nouméa (3.3–6.5 mm/d). The east coast, which is cloudier, records lower values.

The long-term evolution of the climate can already be seen by temperature increases of 1 °C over the historical period (1960–2010). Models (Dutheil, 2018) predict an increase of around 3 °C by the end of the century, slightly higher on the west coast. Precipitation across New Caledonia is expected to decrease by around 20%, mainly during the southern summer. The number of heat waves could double or even triple in the southwest of Grande Terre, and the number of days considered as heat waves could be multiplied by 12. We will therefore see an extension of these episodes and droughts, in addition to an increase in their frequency.

Consequently, even though they are primarily caused by human activities (99%), we should be particularly concerned about both the increase in frequency and magnitude of fires.

2.1.4 Hydrology

The first hydrometric network was installed by ORSTOM from 1954, notably in order to better understand the agricultural and hydroelectric resource. Since 1991, this observation network has been managed by the territorial service responsible for agriculture (currently DAVAR). This network comprises around 40 limnigraphic stations with 21 equipped with remote data transmission systems and 70 pluviographic stations.

These data made it possible to determine typical flow rates for 24 stations within the monitoring network equipped with remote transmission (Alric, 2009; Domet, 2022). In order to be meaningful and comparable, the flow rates are expressed in specific values (l or $m^3/s/km^2$), that is to say normalised according to the surface area of the watershed.

Hydrological patterns are closely linked to rainfall patterns. In accordance with the distribution of precipitation, the annual average specific flow rates are higher (28–55 $L/s/km^2$) on the "windward" slopes of the south and the east coast, while the values are frequently lower than 20 $L/s/km^2$ in the northern half of the west coast (Wotling, 2012).

In addition, rainfall patterns greatly influence infiltration and runoff processes depending on soil saturation. The more abundant the rains, the higher the runoff coefficient. It varies from 20% to 80% depending on the stations. For a given station, it reaches its maximum at the end of the wet season and minimum at the end of the dry season. This means that the month of October generally has the lowest hydraulicity, while the highest average flows are usually observed in March. The wet season represents more than 70% of flows, including more than 50% in the first quarter, while the driest quarter (September–November) only accounts for 6–10% of annual volumes (Wotling, 2012). This seasonality is especially marked in the northwest quarter of Grande Terre while it is least pronounced in the South.

The share of runoff is on average 58% of precipitation on the east coast, compared to 41% on the west coast. The flow deficit ranges on average between 600 and 1200 mm/year depending on the climatic and geological setting (Jeanpert et al. 2021).

The primary feature of surface water resources is essentially characterized by residual flows during periods of low water. The term low flow (LF) refers to the average daily flow rate below which the flow falls 10 days annually. Once again, the major climatic regions of Grande Terre are expressed with LFs often higher than 4 $L/s/km^2$ for the wettest regions of the east coast, while the most severe low water levels occur in the north-west sector (LF <2 $L/s/km^2$), where complete drying up of certain watercourses can be observed.

The far south presents contrasting low flow, due to a particular hydrogeological functioning of the pseudo-karst type. The presence of peridotites within the watershed actually helps sustain river flows during low water periods.

Typical drying times are generally short (1–3 months) and rainwater remains in the watershed for a very limited time. Consequently, the rains from the April to October period have the greatest influence on low flow while those of the first quarter, although more abundant, have almost no effect on low flow patterns (Wotling, 2012).

In New Caledonia, floods stand out primarily by their intensity (most often linked to cyclones and tropical depressions), and secondly for their sudden onset, largely a result of the watershed's limited size and steep gradients. The intensities and accumulations of rain explain soil saturation phenomena and high runoff coefficients, frequently between 40% and 60%, and sometimes exceeding 80%, including for large watersheds. These flows are then quickly concentrated downstream of the watersheds, the water concentration and rise times rarely exceeding a few hours. Hydro-meteorological observations make it possible to estimate transfer speeds of between 1.5 and 5.0 m/s depending on the longitudinal slope, representing an average of around 2.6 m/s.

Given the high rainfall intensities and reduced concentration times, the specific flood flow rates are very high, from 10 to 30 $m^3/s/km^2$ for a 100-year flood (Wotling, 2012). Despite abundant rains, the south stands out for the lowest flood flows, explained by greater concentration times resulting from a particular geomorphology (plateau and lakes) and hydrogeological functioning. The eastern sector has specific flow rates almost twice those of the west coast. This relationship is consistent with that which can be determined on the spatial distribution of centennial rainfall.

The banks of lowland rivers very frequently overflow due to significant differences between flood flow rates and normal patterns. Localised overflows are generally observed from the annual flood and the five-year flood already nearly cover the entire alluvial plain. The highest water levels easily reach 6–8 m above the low water level and levels higher than 10 m can be recorded on most limnigraphs and up to 18 m for the Diahot River (Wotling, 2012).

These assessments of flood flows and recorded flood levels recorded in the floodplains (alluvial plains) act as benchmarks for calibrating hydraulic models that delineate flood zones. While areas subject to flooding can be extensive, submergence durations often remain very limited (rarely more than a few hours) given the rapid response of the watersheds.

2.1.5 *Hydrogeology: Hydrogeological Atlas and Major Hydrosystems*

The BDLISA-NC, hydrogeological atlas (Aquifer Systems Limits Database[2]), is a tool for locating groundwater data, providing mapping and a description of hydrogeological entities (permeable, semi-permeable or impermeable underground geological formation) according to four main characteristics (theme, nature, environment, state) different properties (nature, environment, state, theme) at 1/50,000 (level 3, local). The database consists of unit information relating to hydrogeological entities and their respective characteristics. Representations at smaller scales (level 2 or level 1) are generated through mechanisms combining individual units.

Detailed geological maps and hydrogeological data available facilitated the design (Mardhel et al., 2019; Mardhel & Jeanpert, 2020), creation (Goué et al., 2021a, b) of this database or hydrogeological atlas for New Caledonia (BDLISA-NC, Fig. 2.3). The work initiated by the government of New Caledonia and the BRGM was subsequently financially supported by the European Union within the

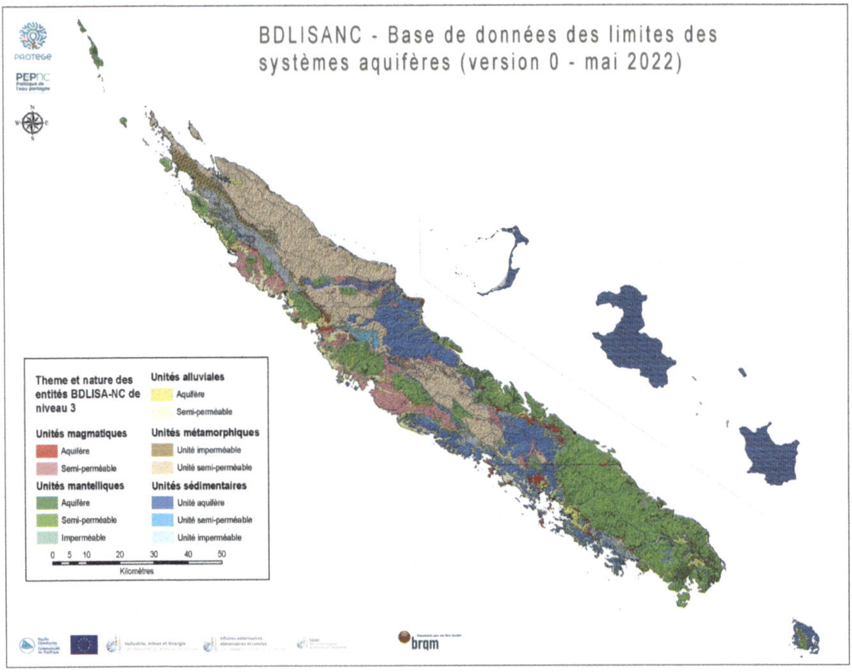

Fig. 2.3 Hydrogeological atlas BDLISA-NC (DIMENC-SGNC, BRGM, 2022)

[2] https://bdlisa.eaufrance.fr/.

framework of PEP.nc and the Oceanian regional project of territories for sustainable management of ecosystems (PROTEGE).

The BDLISA-NC is available online and includes 29 levels 1 entities (approximately 1/500,000), 213 levels 2 entities (approximately 1/200,000) and 469 local levels entities (level 3, 1/50,000).

The territory's main hydrosystems align with the major tectonic and geological domains (Fig. 2.4). Each of them is also outlined using a conceptual vertical layout-including outcrop or underground elements, allowing a quasi-3D approach.

- From the surface to deeper layers, taking into account their overall extent (outcrop and underground), the main hydrosystems are (Fig. 2.5):
- The fluvial and coastal aquifer system which represents only 10% of New Caledonia, only at the outcrop, is the major underground resource used in the territory.
- The carbonate aquifer system of the Loyalty Islands which represents 11% of the territory, only at the outcrop.
- The multilayer system of peridotite massif, present over 31% of the territory, mainly at outcrop (92%).
- The multilayer system of magmatic formations of the Poya unit over 35% of New Caledonia, mainly under cover (84%).
- The hydrogeological domain of sedimentary formations from the Upper Cretaceous to the Eocene, over 48% of the territory, mainly under cover (78%).

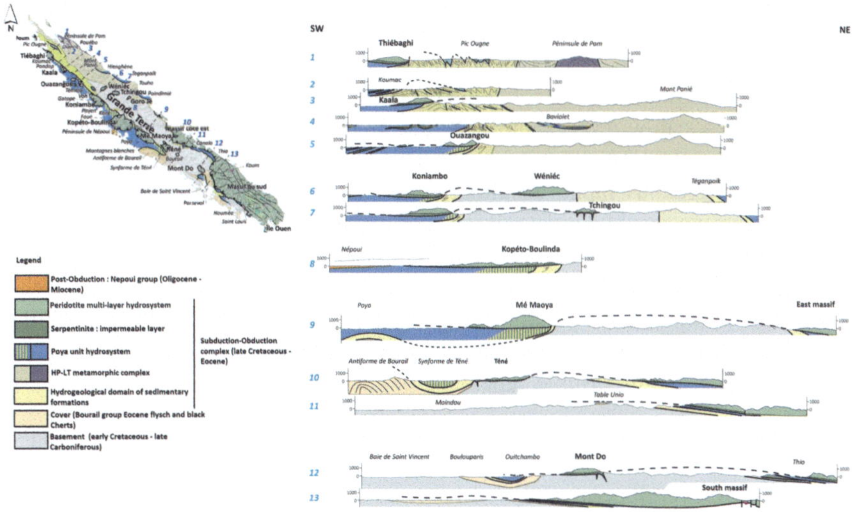

Fig. 2.4 Main hydrosystems according to the major recognized tectonic domains (Maurizot et al., 2020b; DIMENC-SGNC, BRGM, 2023)

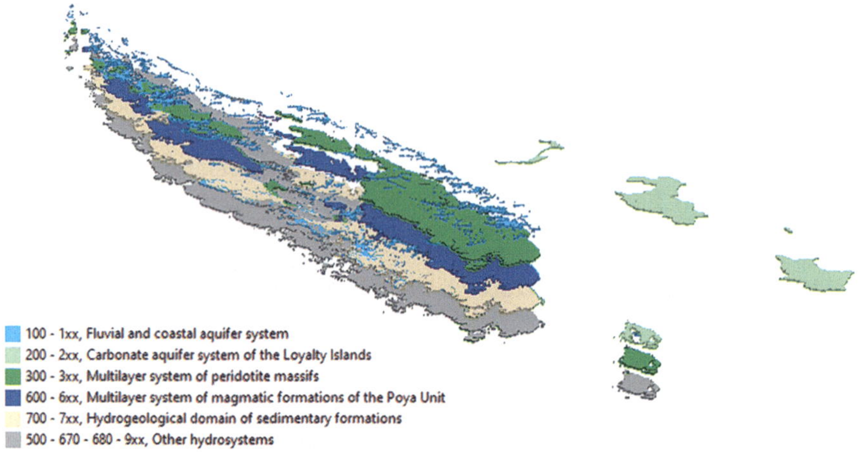

Fig. 2.5 Hierarchy of hydrosystems within BDLISA-NC (DIMENC-SGNC, 2023)

2.1.5.1 The Fluvial and Coastal Aquifer System (Fig. 2.6)

Within the main plains, modern and ancient alluvial formations can be distinguished, with differing lithology.

Given the geomorphology of the territory, lateral extension of alluvial plains is often small, from a few tens of metres to very rarely a few hundred metres. The ancient alluvial formations are mainly clayey and form fairly thin deposits. They are not very conducive to water storage, unlike modern formations, which are more permeable (Espirat, 1967). Modern alluvial formation is generally limited to a strip of several tens of metres to a few kilometers to the right of the alluvial plains, where the watercourses flow. For example, at the mouth of the Tontouta River, where coastal and fluvial sediments accumulate. This is especially the case when there are significant terrigenous inputs resulting from various erosive pressures including fires, invasive species, historical mining activity, and so on.

The nature and grain size of deposits is closely dependent on the watershed's geological formations, their erosion and transport along hydrographic networks. This results in significant spatial heterogeneity, including in terms of permeability (linked to the diameter of the deposits) and porosity. However, it is possible to identify some trends are discernible depending on the nature of the watersheds:

- The Poya unit formations and the finest sedimentary facies from the Upper Cretaceous to the Eocene favour clayey to silty alluvium and therefore low to almost zero overall permeability;
- The ultramafic, carbonate, schist formations and breccia facies of the Flysch produce coarser alluvium and therefore have an average to very good overall permeability (example of the coarse alluvium in Koumac, where the Darcy coefficient has been measured at 10^{-2} m/s).

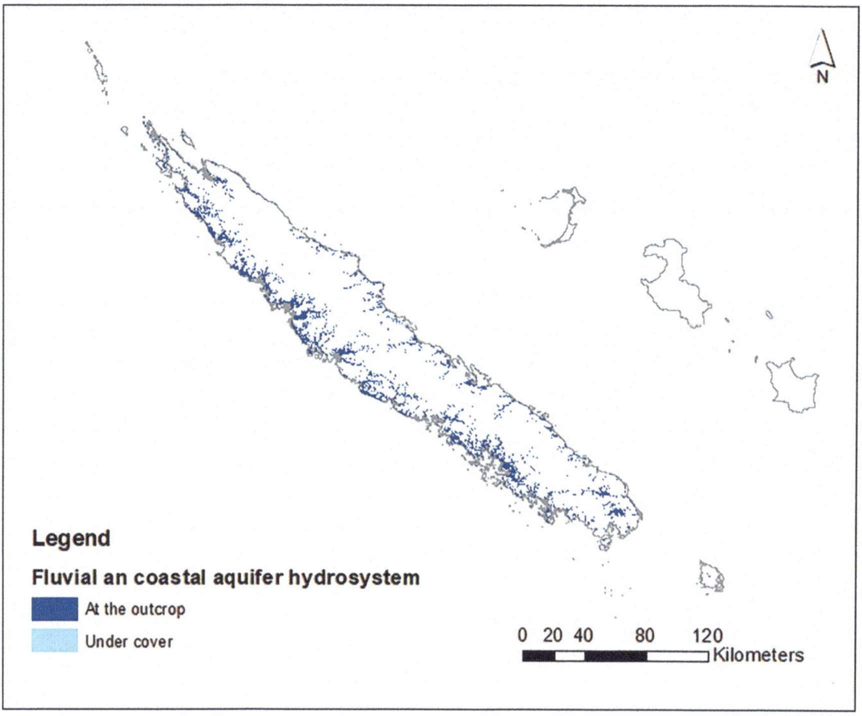

Fig. 2.6 Fluvial and coastal hydrosystem (BdLISA NC)

The large alluvial plains present significant drinking water (Tontouta, Voh for example) and/or agricultural (Pouembout) issues and are equipped with monitoring networks, managed by the DAVAR (Directorate of Veterinary Food and Rural Affairs).

Underground marine intrusion is observed within the hydrosystems which makes water resources very close to the coast vulnerable. With the exception of the Loyalty Islands, the influence of the salt wedge is mainly encountered in the large alluvial plains when the piezometric levels are at low altitudes; levels which depend both on the aquifer's annual recharge but also on the geometry of the reservoir and in particular its bedrock.

The movement of the salt wedge front is much more sensitive to a drop in the fresh water load (reduction in recharge or increase in drainage pressure) than to a change in the salt water load.

The salt wedge is also compounded by saline intrusion where saltwater enters from the surface through specific river mouths during intense high tides. These repetitive marine inputs can locally disrupt the salinity of the aquifer. On the territory, these intrusions were controlled when necessary with regard to upstream issues (Pouembout, Koné, La Foa for example) via the installation of anti-salt fences (small weir) across watercourses.

2.1.5.2 Carbonate Hydrosystem of the Loyalty Islands

The Loyalty Islands (Maré, Lifou and Ouvéa) are uniformly composed of carbonate deposits, mainly rhodolith limestones, locally topped by a dolomitic hardground on Maré (Maurizot et al., 2020c). Coral reef constructions are preserved as raised rims across all three islands, indicated an atoll stage in their development. Unique for Maré, Middle Miocene intra-plate basalts and associated volcaniclastic rocks form restricted outcrops are unique to Maré. Although typically volcanic, the nature of the deep Loyalty Ridge remains unknown.

The hydrogeological setting of the Loyalty Islands comprises carbonate formations, predominantly karst. This phenomenon of karstification finds its origin during the glacio-eustatic oscillations of the Quaternary which led to a succession of phases of emersion and submergence. It is during these phases that the mechanisms of meteoritic alteration of carbonate rocks play a role, promoting the infiltration and circulation of rainwater, via natural fracturing and the presence of primary voids in the carbonates. These phenomena favour the dissolution of carbonates (acid potential of rainwater and relatively high solubility of these rocks) and create karst figures such as conduits, caves, sinkholes, including some underground cavities that can be visited.[3] These underground cavities can constitute significant reserves of freshwater. Moreover, karst aquifers are vulnerable due to the rapid circulation of water. The lack of soil or its thinness, along with exceptionally high permeability of rocks and the weakness of hydraulic gradients expose the freshwater lens to various sources of pollution (infiltration of hydrocarbons, pesticides or leachate from uncontrolled landfills, etc.).

These islands are devoid of perennial watercourses due to the rapid infiltration of meteoric waters, via natural fracturing and the primary voids present in these carbonates. Rainwater accumulates at depth, forming a fresh water lens sits in equilibrium on the denser sea water intrusions ($d = 1$ fresh water and $d = 1.025$ for water of marine origin), along the salt wedge (Fig. 2.7).

The water needs of Maré and Lifou islands are met by pumping from the fresh water lens, a unique resource on these islands. This resource is critical, given the salt wedge and its vulnerability to pollution (Fig. 2.8). Without careful preservation of its quality and quantity, the sustainability of this resource is seriously threatened.

Seawater desalination is sometimes a necessary solution (Ouvéa, Tiga). However, in the Antilles hurricanes Hugo in 1989 or Irma in 2017 hurricanes have demonstrated the fragility of these infrastructures, respectively in Desirade and Saint-Martin islands and the value in having access to an alternative underground resource, even one of low capacity, after a major disaster.

[3] https://www.nouvellecaledonie.travel/offres/les-joyaux-de-luengoni-lifou-fr-3915343/.

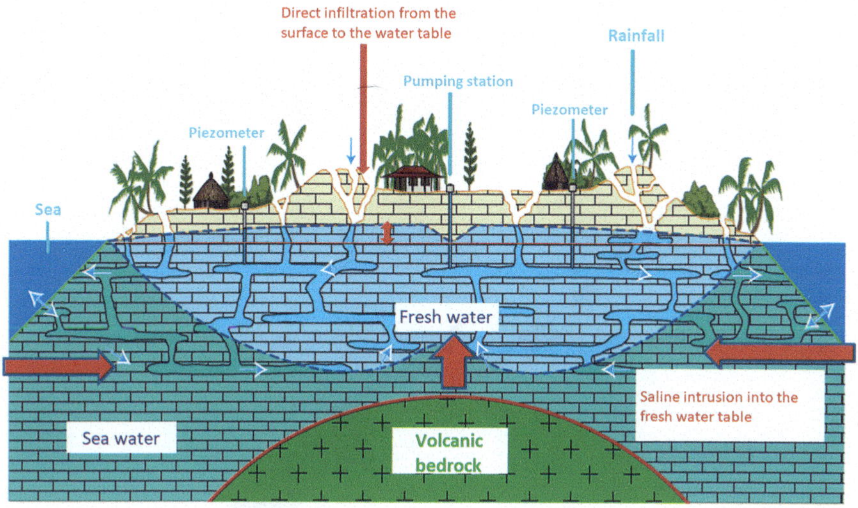

Fig. 2.7 Schematization of the karst aquifer of Lifou island and its vulnerability to surface pollution and marine intrusion (DIMENC-SGNC)

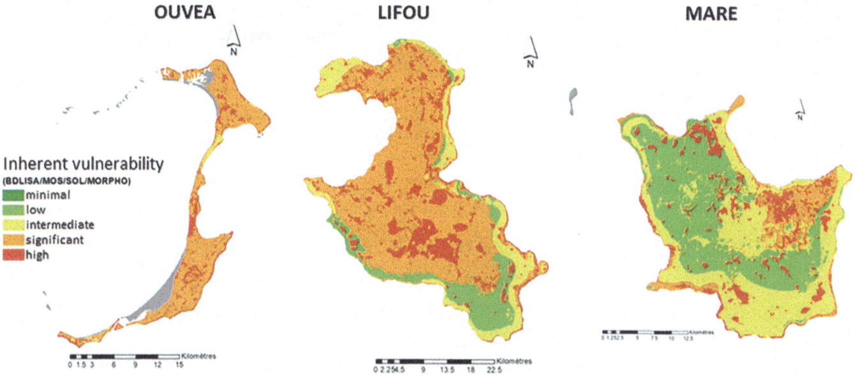

Fig. 2.8 Inherent vulnerability taking into account the unsaturated groundwater zone of the three Loyalty Islands (DIMENC-SGNC, BRGM, 2022)

2.1.5.3 Peridotite Multi-layered Hydrosystem

A third of New Caledonia is covered by an ophiolite nappe, represented by a large massif to the south and several small isolated massifs up to its northern extremity (Fig. 2.9). The peridotite massif and the alteration in its mantle have been studied for more than a century for their nickel ores, but also iron, chrome and cobalt. While research into the mineralisation and formation of deposits, as well as the role of hydrogeology and paleohydrogeology have been ongoing since the 1970s to explain the genesis of rich deposits, the examination of water resources within peridotite massifs has only emerged in the last decade (Join et al., 2005).

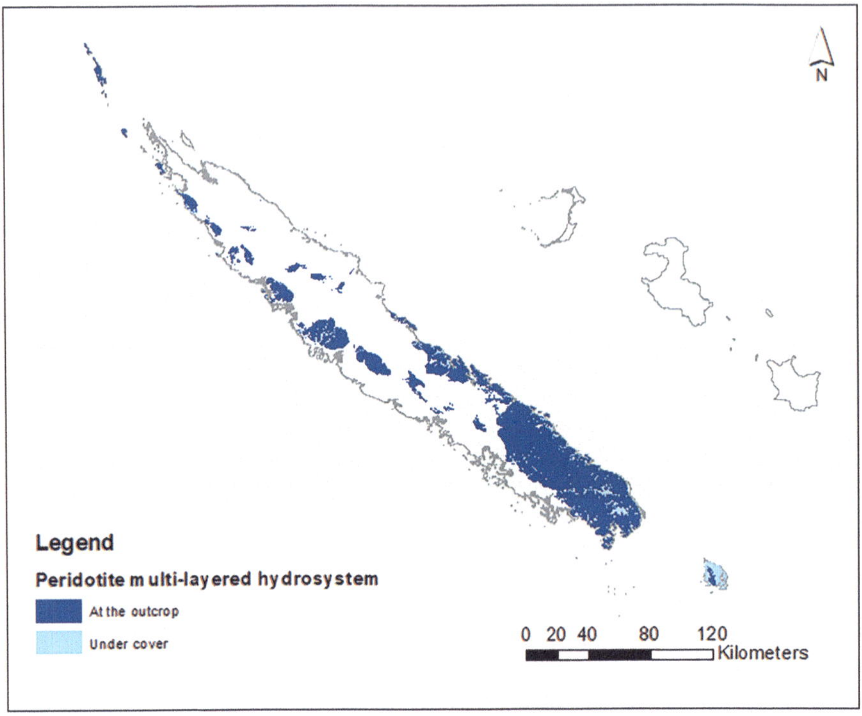

Fig. 2.9 Distribution of the peridotite hydrosystem (BDLISA NC)

The accumulation of work carried out in New Caledonia, particularly on mining hydrogeology and under the leadership of the CNRT (Jeanpert and Dewandel, 2013; Join et al., 2017; Jeanpert et al., 2019; Jeanpert, 2017) has facilitated the characterisation of the hydrosystem within peridotite massifs, conceptualized as a large multilayer system consisting of four hydrodynamically different strata (Fig. 2.10) which resting on serpentinite when the sequence is complete. Indeed, bedrock (deep peridotite) can be compared to a fractured environment, the hydrosystem's structure is highly dependent on the "supergene" alteration profile which can reach several tens of metres under the influence of the tropical climate (Sevin, 2014).

The peridotite hydrosystem has a high level of intrinsic vulnerability[4] (Goué et al., 2022). This vulnerability is that of the first aquifer formations encountered. On the scale of the peridotite hydrosystem (superposition of several aquifers), it is more difficult to define overall intrinsic vulnerability.

[4] Describes whether or no underground reservoirs may be reached by pollutants from the surface of natural lands (qualitative vulnerability of the resource).

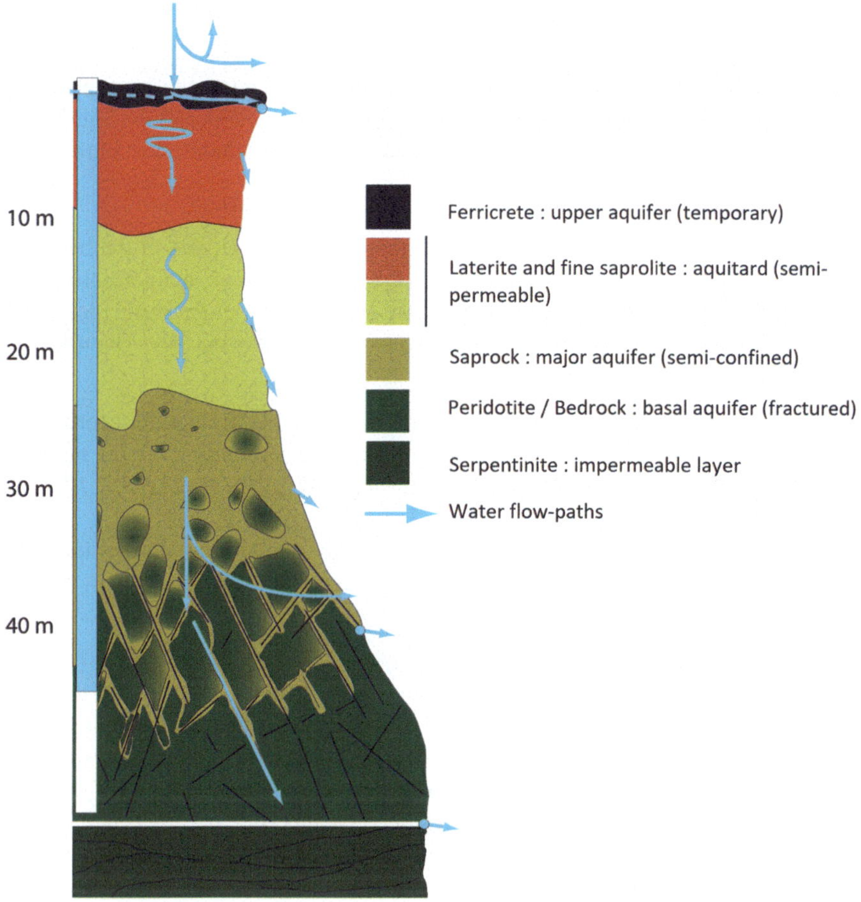

Fig. 2.10 A hydrogeological model of peridotite nappe from New Caledonia (modified figure from the HYPERK project, CNRT, 2017)

- The upper aquifer
  The temporary upper aquifer is identified to the right of the ferricrete (iron cap). It has double permeability (alveolar due to its structure and fracture due to its dismantling) and promotes rapid infiltration of rainwater and run-off water. The strong contrast in permeability with the underlying laterite induces rapid circulation of water, which often gives rise to a line of temporary springs on the edge of caprock plateaus. In the case of significant precipitation, the temporary water table in this aquifer can become charged under the iron cap and locally give rise to artesian springs (Join et al., 2005).
- The aquitard or semi-permeable unit
  This level corresponds to the levels of red and yellow laterites, as well as fine saprolites. It is characterised by having high porosity and low permeability, which contribute to its capacitive function within the system (Join et al., 2005).

It is often described as being comparable to a sponge. Seepage zones can be observed locally on laterite outcrops during periods of high water.

The hydraulic conductivities measured by in situ tests tend to cluster around the value of 10^{-7} m/s, with low variability, typically within one order of magnitude.

This level is in hydraulic connection with the shallow sinkholes most of the time.

- The major semi-confined aquifer

  This aquifer corresponds to the level of the saprock at the roof of the bedrock. The heterogeneous system composed of very weathered rock facilitates the water circulation. This aquifer serves as the transmissive component of the peridotite multi-layer system supplying the main resurgences of the massifs. The majority of water mineralisation also takes places at this level.

  The measured hydraulic conductivities are very variable, from highly permeable to almost impermeable, highlighting significant heterogeneity. They are between 10^{-8} m/s and 10^{-5} m/s, sometimes 10^{-3} m/s and typically sitting around 10^{-6} m/s (Jeanpert, 2017).

  This type of aquifer regularly exhibits karst behaviour with rapid water transfers and underground connections between different watersheds.

- The basal aquifer

  The fractured peridotite serves as a deep aquitard. Seeps observed in the bedrock tend to prove the existence of a deeper underground flow to the right of the fractures (example of the Chromical tunnel in the north). Although no data exists on the state of fracturing within bedrock from New Caledonia, a similar study on peridotites from Oman (Dewandel et al., 2005) suggests an extension of the fracturing zone over several tens of metres deep (Jeanpert et al., 2019) highlighted the importance of altered fractures with enhanced permeability within a micro-fractured matrix with low permeability, representing hydraulic conductivity of around 10^{-8} m/s that diminishes with depth.

  The peridotite hydrosystem sits atop on serpentinite rocks which constitutes an impermeable limit at the base of the massifs and is where a line of springs originates.

- Functioning and karstification

  This hydrosystem contributes to the supply of hydrographic networks and associated aquifers, in particular the underground Poya unit (Fig 2.11).

The peridotite hydrosystem of New Caledonia thus consists of a multilayered system, with each stratum presumed to have distinct hydrodynamic properties. Despite having a structure similar to that of the base aquifers, it has characteristics that make it a specific hydrosystem.

The saturation state of the hydrosystem strongly depends on the geomorphology. Three main types of operation can be distinguished:

- "Stage Plateau" or "Drained and perched" type massif: characterized by deep incisions of "V"-shaped valleys defining interfluves with sharp crests (for example: Koniambo and Kopeto).

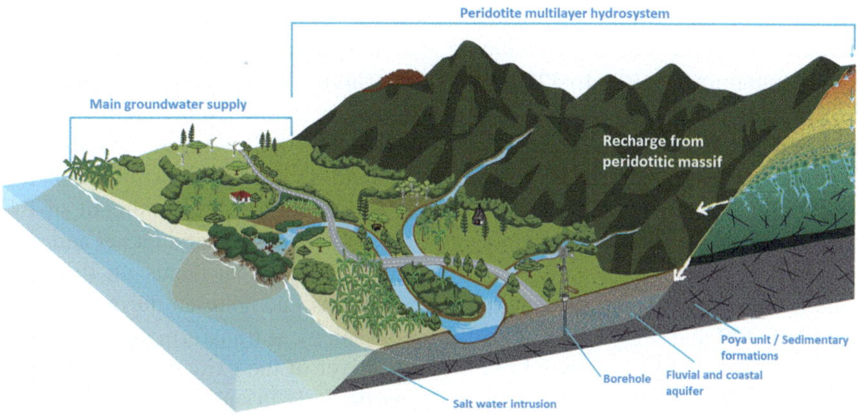

Fig. 2.11 Interaction between the peridotite hydrosystem with others (DIMENC-SGNC)

- "Plateau" or "Poorly drained and perched" type massif: characterised by a flat surface with a slight inclination occupying an area of several km^2 (for example: Tiébaghi, Poum and Bélep).
- "Basin" type massif: characterised by a flat surface with a slight gradient occupying an area of several km^2 bordered by ridges (for example: massifs in the extreme south of Grande Terre).

The solubility of silicates allows the formation of voids, or even drains, as is observed in limestone environments, except that for peridotites the product resulting from hydrolysis generates a solid alteration residue likely to block the voids formed. In the case of peridotites from New Caledonia, we refer to pseudo-karst (Genna et al., 2005; Nicod, 1994). The large peridotitic plateaus are home to numerous closed depressions that can be described as sinkholes (Genna et al., 2005). Lapiaz and other morphologies typical of a karst model are also observed. As the connection with an underground karst network is not systematically demonstrated, and given these rocks are poorly soluble, the altered of by hydrolysis or mechanical erosion, the term "pseudo-karst" is preferred to designate this observed behaviour the peridotite nappe of New Caledonia.

– Geochemistry

The Peridotite nappe provided the country with a wealth of valuable mineral resources, including Ni, Co, Cr and Fe, and contains other prospective resources, such as the Pt group elements and Sc (Maurizot et al., 2020d). In fact, the soils that develop in New Caledonia and the water that circulates there have metal concentrations higher than those measured in other geological contexts. These metallic elements are more or less mobile (soluble) and bioavailable (extractable by plants).

As part of CNRT projects, studies have shown the importance of suspended matter as a vector of metallic elements with total concentrations much higher than those in dissolved matter and more recently the influence of fire on the dynamics of chromium and nickel within the ultrabasic watersheds in New Caledonia (Thery, 2023). A convincing process has been proposed to explain the nickel contamination (up to 4200 µg/L) in a drinking water supply catchment on the Ile des Pins after a fire episode and during a dry period. Note that this collection took place next to a wetland. The fires that occurred favoured the oxidation of sulfide minerals in the soils of the dried wetland. The resulting acidification led to the dissolution of nickel-bearing chrysotile during the impoundment of the sediments, which released high concentrations of magnesium and nickel and also of sulfates into the downstream catchment. Heating experiments on different soils in New Caledonia reproducing the conditions encountered during a forest fire (200, 400 and 600 °C) show that the mineralogical transformation resulting from the action of heating would be a major driver of the oxidation of Cr(III) to Cr(VI), thus facilitating the mobility of Cr, particularly in surface water after rain.

2.1.5.4 Poya Unit Hydrosystem

The Poya unit is a complex of overlapping layers of oceanic crustal material (basalts, dolerites and radiolarites), stripped from the superficial part of the oceanic crust of the South Loyalty Basin and its boundary with the Norfolk Ridge (Maurizot et al., 2020c). This formation is very extensive on the west coast (between Bourail and Koumac) and more sporadic elsewhere (Fig 2.12).

The volcanic formation's terrain is extensive along the west coast, and also sporadically observed on the east coast, in the south (at the base of Mont-Dore for example) and in the far north.

The Poya unit is a multilayered hydrosystem composed of a porous but not very permeable layer corresponding to the weathered layer or regolith and which covers a sometimes captive fractured permeable aquifer. This configuration supports the intrinsic protection of this resource, the potential is underestimated. Pumping tests (Espirat, 1967) indicate a permeability of between 10^{-3} m/s and 10^{-4} m/s.

The Poya unit is set up at the front of the peridotite layer, which covers it almost completely. Given this superposition, the Poya unit hydrosystem is the main outlet of the peridotite water table hydrosystem and therefore experiences significant annual replenishment irrespective of meteorological conditions. The Poya unit hydrosystem is therefore a strategic water resource, especially considering its low vulnerability and its significant expansion on the west coast in particular.

The Poya unit hydrosystem has low vulnerability because it is relatively protected by the presence of a low permeability regolith. With its hydraulic characteristics, it is probably the water resource with the greatest potential, identified by Espirat, has yet to be developed.

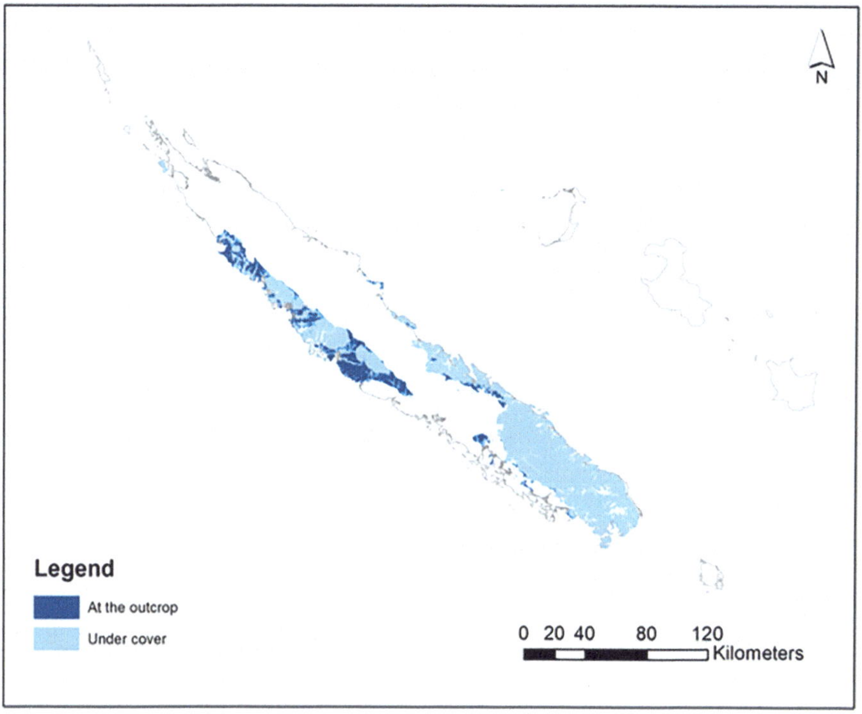

Fig. 2.12 Poya unit hydrosystem (source : BdLISA NC)

2.1.5.5 The Sedimentary Formations Hydrosystem (Upper Cretaceous to Eocene)

Cretaceous sedimentary formations bring together varied lithologies: siliceous shales, coal formations, sandstones, volcanic levels and black claystones and silts, while Paleocene facies have a significant carbonate composition.

Overall, these rocks form poor aquifers because they are not very permeable. Often altered on the surface, the porosity of regolith allows for a small amount of water storage, capable of supporting minimal rivers levels provided that the water supply basin drains sufficient rain. The hydrosystem of sedimentary formations therefore contains aquifer units without much lateral extension that are relatively isolated (Fig. 2.13). Yet, it is crucial to optimise the use of this weak and disperse resource to provide an alternative solution to the distribution of bottles or the delivery of water by tanker trucks for municipalities affected annually by a seasonal rainfall deficit and marked and with a very intermittent hydrographic network (the Poum commune for example). However, the priority solution in these sectors is to optimize the performance of distribution networks (leak detection, installation of buffer tanks, raising awareness among the population, etc.).

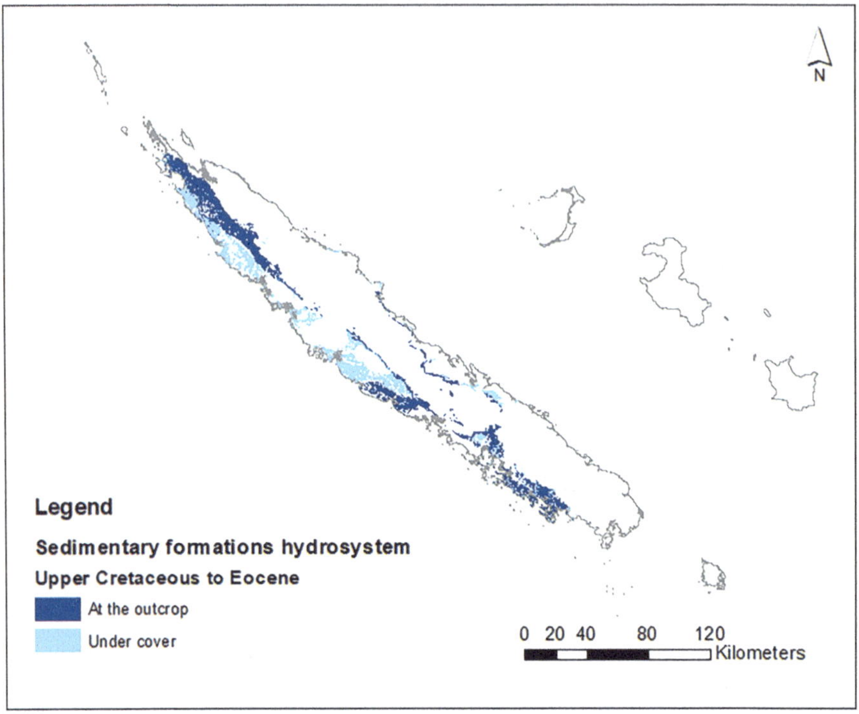

Fig. 2.13 Sedimentary formations hydrosystem (source: BdLISA NC)

2.2 Key Facts and Issues Concerning Water Resources in New Caledonia

There are several constraints associated with context of New Caledonia's specific context (orientation plan for a shared water policy in New Caledonia, PEPnc, 2019):

- The averages taken, although abundant, are very uneven in space and time and regularly expose the territory to episodes of drought, flooding or trigger land movements.
- Water use is separated from resources: "we need water most where there is little and when there is least!" To rebalance uses and resources, water must be stored and/or distributed through major and costly developments, such as: the Dumbéa dam, the large Tontouta pipes or the Pouembo dam project.
- Human activities (industrial, mining, agricultural, etc.), forest fires and the proliferation of invasive species lead to an accentuation of leaching and erosion phenomena generating an increase in solid transport in downstream watercourses. The terrigenous particles coming from these bare surfaces and arriving in aquatic environments are naturally rich in metallic elements (mainly: Ni, Co, Cr, Mn and Fe) and contribute to increasing the natural metal load in the waters.

The abiotic or biotic environmental compartments of waterways can then be impacted.

- On Grande Terre, more than 60% of the resources used for drinking water supply come from surface waters (compared to only 5% in France), even though they are the most vulnerable to pollution, floods and droughts. According to existing estimates, surface water probably represents more over 90% of the volumes exploited.
- Groundwater presents water of better quality and in more constant quantity. On Grande Terre, this resource remains largely unrecognised and underutilized. In the Loyalty Islands, freshwater lens is the only available resource (apart from Tiga and Ouvéa) but it is vulnerable to pollution and saline intrusion.

The main issues to consider with regard to water resources are:

- Define and identify strategic surface and underground resources.
- Improve knowledge to better protect, preserve, plan, manage.
- Raise awareness and communicate in order to limit conflicts of use.
- Put water at the centre of development and energy transition projects.

2.2.1 Define and Identify Strategic Surface and Underground Resources

Protecting catchment areas, defining strategic resources and preserving environments is the first strategic objective (OS1) of the Shared Water Policy (PEPnc). In this context, so-called "strategic" resources were initially defined as ones that, if they were to deteriorate or disappear, would seriously compromise, without any possible alternative, the living conditions of populations, the proper functioning of ecosystems or the maintenance of economic activities considered priorities.

Taking into account the duality between current tense situations and promising resources, a distinction could be made between:

- Critical water resources: low redundancy, high demand, threatened, vulnerable, low residual productivity, etc.;
- Future or strategic water resources: significant potential, naturally protected or slightly vulnerable, etc.

It appeared relevant to develop the means of diagnosing the issues and threats or pressures which may concern a water resource the nature of which could vary depending on the uses and interests of the water resource (drinking, economic, ecological, or heritage water). To do this, the government of New Caledonia and the BRGM, financially supported by the European Union within the framework of the PEPnc and the Oceanian regional project of territories for the sustainable management of ecosystems (PROTEGE), carried out work to develop of hydrographic and hydrogeological benchmarks by offering tools to help determine strategic water resources (Goué et al., 2022). These tools can be configured which remain configurable and assessments can be replicated to take into account new data or the evolution of knowledge.

The modelled hydrological reference frame (MHR1ha) allows the inventory of pressures upstream of a point and to estimate their consequences towards downstream issues by adapting different propagation rules. The nature of the pressures varies greatly; they must be standardised according to their geometry. This may include bare soil, built-up areas and even water withdrawals, for example.

The hydrogeological atlas (BDLISA-NC) was enhanced by incorporating AEP pressures and issues and producing a map of intrinsic vulnerability. The method used combines and assesses four criteria: the characteristics of the hydrogeological entities (state, environment, nature), land use, lithology and geomorphology. The evaluation is further modulated according to the total rain received, land gradient, characteristic low flow rates of the elementary watersheds and the assumed thickness of the unsaturated land.

2.2.2 *Improve Knowledge to Better Protect, Preserve, Plan, Manage*

The qualitative and quantitative challenges of groundwater make them strategic resources with potential that remains largely to be clarified and developed. A better balance needs to be found between the use of surface water and groundwater.

Improving knowledge of water resources is therefore a major issue in New Caledonia's shared water policy (PEPnc, cross-cutting objective B). Relevant information is needed, updated and shared between managers but also with users.

While the potential of groundwater resources has been identified, it remains to be acknowledged, quantified and effectively utilised. The level of knowledge of groundwater reservoirs depends on the density of available investigation points, whether direct: drilling or work in the subsoil or indirect through the acquisition of data (geophysical studies by example).

Around 5000 points are referenced in the Subsoil Database (BDSS-NC) of New Caledonia but only 1400 structures (28%) include the measurement, often unique, of an underground water level (Fig. 2.14). This database does not include all the mining prospecting drillings, nor the existing piezometers on the mining sites. The information density is on average 0.075 water points per km^2 (12 times less than in France). The map below also highlights a disparity in the distribution of points listed.

Furthermore, the spatial distribution of knowledge that can be used for water resources is conditioned by land use and other uses (agricultural, drinking water supply, industry, etc.). Water levels are more frequently known near alluvial layers as they are monitored by government departments, in the Loyalty Islands (vulnerability of the freshwater lens), but also near mining sites and quarries (regulatory

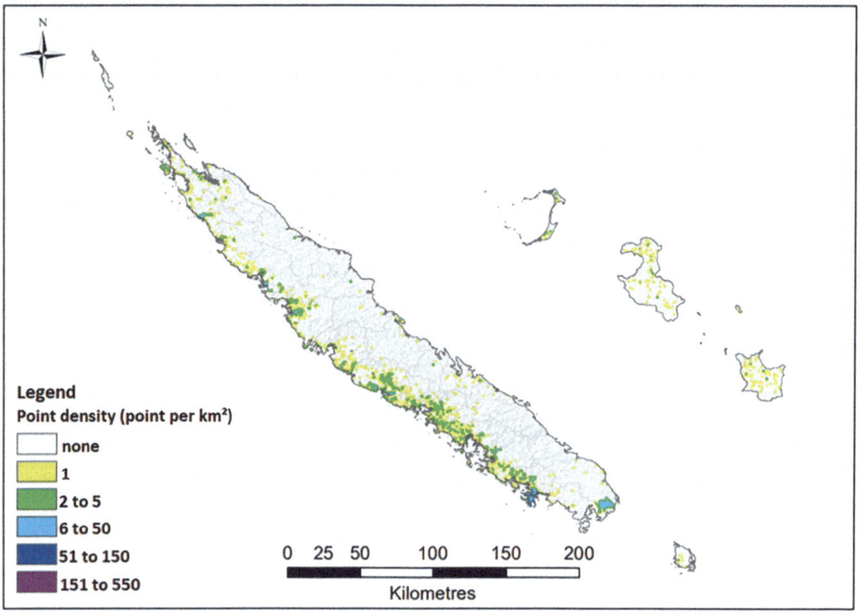

Fig. 2.14 Point density listed within the BDSS-NC (DIMENC-SGNC, BRGM)

requirement for recognition hydrogeological and monitoring). Geophysical data are much more numerous near the peridotite masses on which mining concessions exist.

The density of geophysical information available is significantly greater although more difficult to estimate: different methods and campaigns have been carried out for several decades, whether for private individuals in the case of drilling or mining companies in order to survey their land or even, authorities as part of a study on the location of the salt wedge on the west coast for example. Acquisitions of helicopter-borne electromagnetic data (EMh) financed by the National Center for Technological Research, CNRT, (OPHIOSTRUCT project, 2018), local authorities or the European Union (PROTEGE project in the Poum commune, 2019) have demonstrated the value of this technique, particularly for hydrogeology.

Older data (1980/1990, BRGM) is available for Grande Terre but must be archived and assessed: this includes magnetism campaigns and geochemical analysis of sediments, conducted as part of an inventory for mining alternative substances to the extraction of peridotite masses (other than Nickel, Chrome and Cobalt).

Physical and chemical analysis are fundamental in terms of water quality, in particular to assess potability and determine an average geochemical profile, and identify spatial and temporal variations (droughts, fires, heavy rains). Despite several past and current scientific projects (CNRT, CRESICA, OFB), and lack of sufficient data, the geochemical background remains to be defined in New Caledonia.

The main areas for improving knowledge relating to water resources in New Caledonia are as follows:

- The meteorological observation network seems quite operational. The main pitfall being the absence of weather radar in northern Grand-Terre, pending the replacement of the one damaged by cyclone OMA (2019) and the renewal of those in Lifou and Nouméa.
- The hydrological observation network seems quite effective, although it does not cover the entire territory. Data also remains to be acquired to define a benchmark for watercourses in New Caledonia like the European benchmarks in order to understand the concept of water masses locally.
- Data remains to be acquired in terms of knowledge of underground reservoirs and physico-chemical analysis of water.
- The archiving and assessment of soil and subsoil data (BDSS-NC, BDLISA, New Caledonia geological repository) is one of the missions of the territory's geological service. These missions are currently constrained by the difficulties of retrieving data from the various producers and departments, it will therefore be necessary to improve communication in this respect. Updating these databases will make it possible to consider the establishment of a water quality monitoring network representative of the main hydrogeological entities.

2.2.3 Raise Awareness and Communicate in Order to Limit Conflicts of Use

Poum is a commune located in the far north of Grande Terre. Despite a favourable average rainfall per inhabitant, the drinking water supply is very problematic. This part of the territory experiences the most marked seasonality and the most severe low water levels. The catchments are essentially superficial, particularly around the Poum massif, and the dispersed habitat complicates the connection of distribution networks.

The subsoil is mainly made up of sedimentary formations (Upper Cretaceous to Paleocene), having undergone increasing metamorphism towards the east. To the west, the peninsula corresponds to an isolated massif of peridotites. Note that this was part of the exchange of deposits, as part of the mining prerequisite which enabled the Nouméa agreement (1998). It is both exploited as a nickel mine and makes up the impluvium supporting the aquifer of several superficial catchments on its periphery which supply the village of Poum:

- The watershed upstream of the Ponvio catchment is the location of some mining activity.
- Excluding the influence of mining, the Titch catchment is regularly unfit for consumption after heavy rains because of the red water which results from natural erosion.

This results in a typical usage conflict. A resident of Poum can aspire to work with the mine and access 150 L of good quality drinking water per day. Two regulations come into conflict here, the authorisation to exploit the mine and the establishment of a water protection perimeter. This situation led to the search for alternative resources.

Geological field observations (2019), followed by the mining company's airborne acquisition of electromagnetic data over the massif and its perimeters (2019–2020) made it possible to clarify the geological and hydrogeological context in this area (Fig. 2.15). The serpentinite has been identified and mapped and the extension limits of the ultramafic formations have been redefined. The formation of the Poya unit has been observed for the first time in the field and its depth approximated from geophysical data. The hydrogeological model has therefore been redefined, including the multilayered hydrosystems of the peridotite overlying that of the Poya unit, sitting atop or adjacent to the collection of sedimentary formations (Upper Cretaceous to Paleocene).

The extension of the Poya unit hydrosystem outside the influence of the peridotitic massif is small but its position at the foot of the massif protects the aquifer it contains, much better than the surface waters of the creeks flowing over the sides of the massif. It is topped and partly protected by largely impermeable altered formations composed by colluvium from the massif and its own regolith, not very permeable, which partly protects it. Ultimately, the base of the peridotite massif forms the boundary of the continuous water source for the underlying fractured basalts. Six exploratory surveys were carried out in the Poya formations via EMh, five of them present water potential that the municipality must identify and utilise in order to secure its drinking water supply network.

Outside of the Poum peninsula, the sedimentary formations (Upper Cretaceous to Paleocene) are not or only slightly aquiferous. Nevertheless, at the end of 2020, the acquisition of more than 600 km of helicopter-borne electromagnetic data (funding from the European Union) as part of the PROTEGE project and the PEPnc and an in-depth hydrogeological study, made it possible to identify five areas of interest across the commune. These sectors were the subject of a field diagnosis in 2023 and

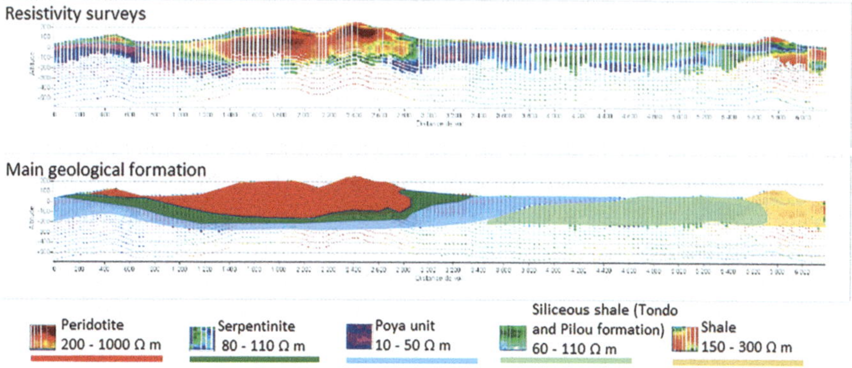

Fig. 2.15 Contributions of helicopter-borne electromagnetism in the geological interpretation of the Poum massif (BRGM, 2023)

the next stage (2024) is a campaign of exploratory surveys to confirm the potential of these small underground resources whose aggregation and connection could finally allow the security of the networks in place and therefore enable all residents to access drinking water in sufficient quantity.

2.2.4 Put Water at the Centre of Development and Energy Transition Projects

The energy transition plan (STENC), approved in 2016 and revised in 2023, sets out the ambition to reduce greenhouse gas emissions by 70% by 2035, in particular by shutting down coal and heavy fuel oil power plants.

One of the associated tasks involves the creation of a countrywide atlas of hydroelectric potential, but not finalised to date, to evaluate and map opportunities and constraints using a multi-criteria approach. Hydroelectricity in New Caledonia has mainly been developed on the east coast with 2 dams and 2 run-of-river power stations which combine favourable rainfall patterns and topography. The annual output is around 345 GWh for an installed power of 80.8 MW. With a volume of 310 million m^3 and a drop of 157 m, the Yaté dam represents 87% of this production. This is the main renewable energy in New Caledonia, accounting for around 17% of the country's energy needs, public distribution and industrial use. With 3 electro-intensive metallurgical plants and increasing photovoltaic production, hydroelectric energy is of interest as it is largely controllable. To meet the need for network regulation, in particular with the development of intermittent energies, a pumped energy transfer station project is seriously being considered in the Tontouta region with a power of 100 MW and a daily capacity of turbines for 9 h or 900 MWh.

The geothermal potential of Grande Terre has been evaluated (Sanjuan et al., 2020) making it possible to better characterise the two very different types of thermal waters:

- The sodium bicarbonate waters of the Thio-Canala region, which release hydrogen sulphide, nitrogen and a little helium of crustal origin.
- The hyperalkaline waters of the south, which emerge from the peridotites releasing natural hydrogen and methane.

The study confirms the absence of indications of high temperature geothermal resources. Given the restricted range of temperatures (from 24 to 42.5 °C on the surface, from 50 to 85 °C with deep drilling of 1–3 km), applications other than balneotherapy/thermalism seem limited.

The proven existence of a natural hydrogen system associated with Peridotite nappe is an opportunity for the territory. It remains to be explored to better understand the processes involved in water circulation and serpentinization and integrate it into the energy mix if necessary. Furthermore, the Poya and Koh units, corresponding to oceanic crust, are also candidates that should be evaluated. Many research study projects are currently being launched or seeking funding on this topic.

2.3 Conclusion

Natural environments depend, among other things, on the presence or absence of water and whether it is fresh, brackish or marine. Aquatic environments (rivers, lakes, aquifers, sinkholes, wetlands, mangroves, etc.) are exposed to multiple development factors which can constitute threats likely to degrade water quality, harm their biodiversity and ecosystem health, in particular those of the lagoon which is where terrestrial waters on the New Caledonian territory culminate.

We have shown that despite an annual rainfall of around 1680 mm, the high variability of rainfall subjects New Caledonia to severe low water levels and therefore to a recurring risk of imbalance between resources and uses. Indeed, many rivers experience total drying up after 90 days without rain, resulting in water resources and aquatic environments being highly dependent on weather conditions, while more than 60% of resources exploited for domestic drinking water comes from surface catchments, which makes them very vulnerable. Added to this is the fact that the way in which the climate is likely to evolve in New Caledonia is still poorly understood. Météo France does, however, envisage more frequent, longer and more intense droughts. Consequently, periods of severe low water which can lead to water shortages in surface catchments are to be studied, as is an increased risk of forest fires. However, fires do also promote soil erosion, the release of suspended matter and trace metal elements.

Since 2019, the recognition of groundwater potential is a key focus of the PEPnc framework for drinking water uses. The first work carried out since the BDLISA NC and other data acquisition campaigns have shown a significant groundwater potential of certain hydrosystems (especially the Poya unit hydrosystem). More broadly, gaining understanding of local hydrosystems is therefore a major challenge to be met in order to secure the water supply of the population, and ensure better preservation of surface ecosystems (reduction of pressure on hydrographic networks).

Acknowledgments This work is part of the missions and public policies of the Government of New Caledonia. It has benefited from the technical and financial partnership of BRGM. It also benefited from a financial contribution from the Pacific Community (SPC) under the PROTEGE programme (Pacific Territories Regional Project for Sustainable Ecosystem Management, financed by the European Union) for the development of the hydrogeological reference framework (BDLISA-NC) and support for the definition of strategic water resources (OS1 of the PEPnc).

References

Alric, R. (2009). *Recueil des débits caractéristiques de la Nouvelle-Calédonie* (p. 314). DAVAR Observatoire de la ressource en eau.

Bonvallot J. (2012) Les milieux (II), Planche 3 L'orohydrographie. Alas de la Nouvelle-Calédonie, Marseille-Nouméa: IRD-congrès de la Nouvelle-Calédonie, pp. 25-28

Bonvallot J., Gay, J-C., & et Habert É. (coord.) (2012). Atlas de la Nouvelle-Calédonie, IRD-congrès de la Nouvelle-Calédonie, Marseille-Nouméa, 272 p.

Dewandel, B., Lachassagne, P., Bourdier, F., Al-Hattali, S., Ladouche, B., Pinault, J.-L., & Al-Suleimani, Z. (2005). A conceptual hydrogeological model of ophiolite hard-rock aquifers in Oman based on a multiscale and a multidisciplinary approach. *Hydrogeology Journal, 13*, 708–726.

Domet, M.-A. (2022). *Recueil des débits caractéristiques de la Nouvelle-Calédonie – Actualisation jusqu'à la période 2019/2021* (p. 405). DAVAR Service de l'eau.

Dutheil, C. (2018). *Impacts du changement climatique dans le Pacifique Sud à différentes échelles: Précipitations, cyclones, extrêmes*. Météorologie (Sorbonne Université. Français. NNT: 2018SORUS402. tel-02468810v2)

Espirat, J. J. (1967). *Etude hydrogéologique de la côte ouest de la Nouvelle-Calédonie*. Rapport d'ensemble sur les travaux effectués par le BRGM de 1962 à 1967. p. 178

Genna, A., Bailly, L., Lafoy, Y., & Augé, T. (2005). Les karsts latéritiques de Nouvelle-Calédonie. *Karstologia, 45–46*, 19–28.

Goué, A., Jeanpert, J., & Mardhel, V. (2021a). *Référentiel hydrogéologique de la Nouvelle-Calédonie BDLISA-NC—Rapport d'étape n°2* – L'atlas hydrogéologique de la Nouvelle-Calédonie (base de données SIG, principes de cartographie hydrogéologique et outils). Rapport DIMENC/SGNC-2021(13), 31 p. Rapport DIMENC/SGNC-2021(05), 50 p.

Goué, A., Jeanpert, J., & Mardhel, V. (2021b). *Référentiel hydrogéologique de la Nouvelle-Calédonie BDLISA-NC - Rapport d'étape n°2* – L'atlas hydrogéologique de la Nouvelle-Calédonie (base de données SIG, principes de cartographie hydrogéologique et outils). Rapport DIMENC/SGNC-2021(13), 31 p.

Goué, A., Mardhel, V., & Monge, O. (2022). *Référentiel hydrogéologique de la Nouvelle-Calédonie BDLISA-NC-Rapport d'étape n°4—Les ressources stratégiques en eau superficielle (Référentiel hydrographique et masses d'eau)*. Rapport DIMENC/SGNC-2022(08). 58 pages.

Jeanpert, J. (2017). *Structure et fonctionnement hydrogéologiques des massifs de péridotites de Nouvelle-Calédonie*. PhD Thesis. Université de la Réunion.

Jeanpert, J., & Dewandel, B. (2013). Analyse préliminaire des données hydrogéologiques du massif du Koniambo, Nouvelle-Calédonie - SGNC/DIMENC. BRGM/RP-61765-FR 95 p., 38 fig., 7 tabl., 2 ann.

Jeanpert, J., Iseppi, M., Adler, P. M., Genthon, P., Sevin, B., Thovert, J.-F., Dewandel, B., & Join, J.-L. (2019). Fracture controlled permeability of ultramafic basement aquifers. *Inferences from the Koniambo massif, New Caledonia Engineering Geology, 256*, 67–83. https://doi.org/10.1016/j.enggeo.2019.05.006

Jeanpert, J., Join, J.-L., Gunkel-Grillon, P., Monge, O., Bouard, S., Lejars, C., & Sabinot, C. (2021). Ressources en eau dans les massifs de péridotites: état des connaissances et enjeux. *Revue Géologues, 209*, 66–73.

Join, J.-L., Robineau, B., Ambrosi, J.-P., Costis, C., & Colin, F. (2005). Système hydrogéologique d'un massif minier ultrabasique de Nouvelle-Calédonie. *Comptes-Rendus Géoscience, 337*, 9.

Join, J.-L., Bouichet, G., Jeanpert, J., Dewandel, B., Genthon, P., Maréchal, J.-C., Adler, P.-M., Maurizot, P., & Sevin, B. (2017). Guide pour l'analyse et le suivi des circulations d'eau souterraine en milieu minier. Programme « HyperK ». CNRT « Nickel & son environnement ». 61 p.

Maitrepierre, L. (2012). *Les milieux (II), Planches 10 et 11 Les types de temps et les cyclones, les éléments du climat* (pp. 53–60). IRD-congrès de la Nouvelle-Calédonie.

Mardhel, V., & Jeanpert, J. (2020). *BDLISA-NC: Construction du référentiel hydrogéologique à partir de modèles de données géologiques adaptés et d'outils cartographiques dédiés*. Rapport BRGM/RP-70134-FR, 42 p.

Mardhel, V., Jeanpert, J., & Balayre, S. (2019). Référentiel Hydrogéologique de Nouvelle-Calédonie BDLISA-NC—Version 0. Rapport BRGM/RP-68516-FR Mai 2019, 42 p.

Maurizot, P., & Mortimer, N. (Eds.). (2020). *New Caledonia: Geology, geodynamic evolution and mineral resources* (Vol. 51). Geological Society, London, Memoirs. https://doi.org/10.1144/M51

Maurizot, P., Robineau, B., Vendé-Leclerc, M., & Cluzel, D. (2020a). *Introduction to New Caledonia: Geology, geodynamic evolution and mineral resources* (Vol. 51, pp. 1–12).

Geological Society of London. Collection. London, Memoirs. https://doi.org/10.1144/m51-2019-33

Maurizot, P., Cluzel, D., et al. (2020b). *The Eocene Subduction–Obduction Complex of New Caledonia* (Vol. 51, pp. 93–130). Geological Society, London, Memoirs. https://doi.org/10.1144/M51-2018-70

Maurizot, P., Collot, J., Cluzel, D., & Patriat, M. (2020c). *The Loyalty Islands and ridge, New Caledonia* (Vol. 51, pp. 131–145). Geological Society, London, Memoirs. https://doi.org/10.1144/M51-2017-24

Maurizot, P., Sevin, B., Lesimple, S., Bailly, L., Iseppi, M., & Robineau, B. (2020d). *Mineral resources, prospectivity of the ultramafic rocks of New Caledonia* (Vol. 51, pp. 247–277). Geological Society, London, Memoirs. https://doi.org/10.1144/M51-2016-17

Mortimer, N., Campbell, H., et al. (2017). Zealandia: Earth's hidden continent. *GSA Today, 27*, 27–35. https://doi.org/10.1130/GSATG321A.1

Nicod, J. (1994). Plateaux karstiques sous couverture en France. *Annales de Géographie t, 103*(576), 170–194.

Romieux, N. (2011). *Synthèse et régionalisation des données pluviométriques de la Nouvelle-Calédonie* (p. 82). DAVAR Observatoire de la ressource en eau.

Sanjuan, B., Jeanpert, J., Lesimple, S., Darnet, M., & Jacob, Th. (2020). *Evaluation du potentiel géothermique de la Grande Terre (Nouvelle-Calédonie). Phase 2: Interprétation des résultats et recommandations*. Rapport final BRGM/RP-69704-FR, 56 p.

Sevin, B. (2014). *Cartographie du régolithe sur formation ultrabasique de Nouvelle-Calédonie: localisation dans l'espace et le temps des gisements nickélifères*. PhD Thesis, Université de Nouvelle-Calédonie.

Thery, G. (2023). *Influence du feu sur la dynamique du chrome et du nickel à l'échelle des bassins versants ultrabasiques de Nouvelle-Calédonie. Milieux et Changements globaux*. PhD Thesis, Université Paris-Saclay.

Wotling, G. (2012). *Les milieux (II), Planche 12 L'Hydrologie* (pp. 61–64). IRD-congrès de la Nouvelle-Calédonie.

Chapter 3
The Balance Between Water Resources and Uses

Pearl Winchester, Geoffroy Wotling, Gérard Fallon, and Aude Arrighi

Abstract In New Caledonia, the balance between water resources and uses is becoming precarious. On one hand, there is pressure from withdrawals to meet the domestic and economic needs of the population. On the other hand, there are "aggressions" that can damage natural areas and alter the quality of water and aquatic environments. The main pressures are terrigenous inputs, such as red water and sedimentation due to erosion (related to fires, mining activity, land clearing, proliferation of deer, etc.) and wastewater discharges from urban, industrial and agricultural origin. This chapter maps water-related demand and uses from different sectors (metallurgic industry, agriculture and domestic) and territories, while highlighting the main regional differences. It then analyses the various pressures that impact the landscapes and water degradation: mining activities, rising logging and the increase of irrigated agriculture development. It also underlines the issues linked to the introduction of invasive species and bush fires. Lastly, it highlights the major drivers of change that are likely to shape and increase water pressures, such as climate change.

Keywords Water demand · Water uses · Tensions and pressures on water

3.1 Context: Distribution of Resources, Demography and Uses

New Caledonia experiences an annual rainfall of approximately 1850 mm, with notable variations observed over a couple of years. Precipitation is concentrated in the first trimester, accounting for roughly 50% of the total, and decreases significantly in the last trimester, representing only about 10% of the annual accumulation.

P. Winchester (✉) · G. Wotling · G. Fallon · A. Arrighi
DAVAR, Directorate of Veterinary Food and Rural Affairs, Water Division, Nouméa, New Caledonia

C. Lejars, S. Bouard (eds.), *Water Policy in New Caledonia*, Global Issues in Water Policy 32, https://doi.org/10.1007/978-3-031-88196-1_3

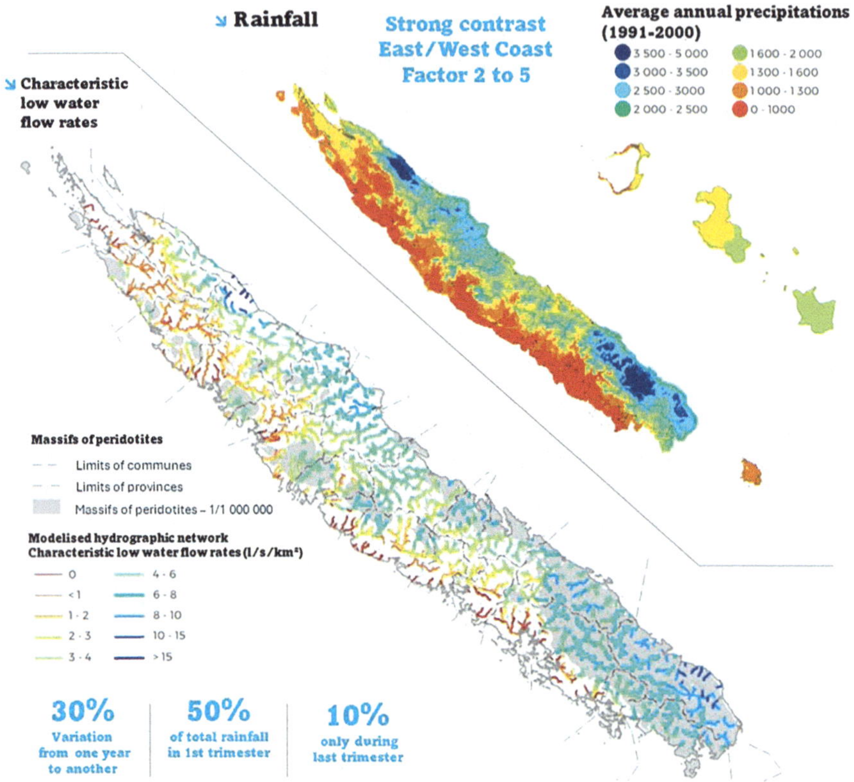

Fig. 3.1 Pluviometry and flow repartition in New Caledonia

Also, as shown in Fig. 3.1, the rainfall regime is very heterogeneous across the territory of New Caledonia. The southern sector, the east coast and the Loyalty Islands receive nearly twice as much rainfall overall than the northern and western sectors of the main island.

This rainfall occurs over New Caledonia's 18,575 km^2 territory. New Caledonia has 268,500 inhabitants (ISEE, 2023) and one of the lowest population densities of all the French overseas regions, after French Guyana, in South America. The average population density in New Caledonia is 15 inhabitants per km^2, with significant differences between the North Province (5 inhabitants/km^2), the Loyalty Islands Province (9 inhabitants/km^2) and the South Province (26 inhabitants/r km^2). The Greater Nouméa agglomeration alone concentrates nearly 70% of the Caledonian population, with an average density of 545 inhabitants per km^2 in the 4 communes of Nouméa, Dumbéa, Païta and Mont-Dore.

Contrasting average annual rainfall volumes with population data by municipality offers valuable insight into the disconnection between rainfall distribution and domestic water demands (Fig. 3.2).

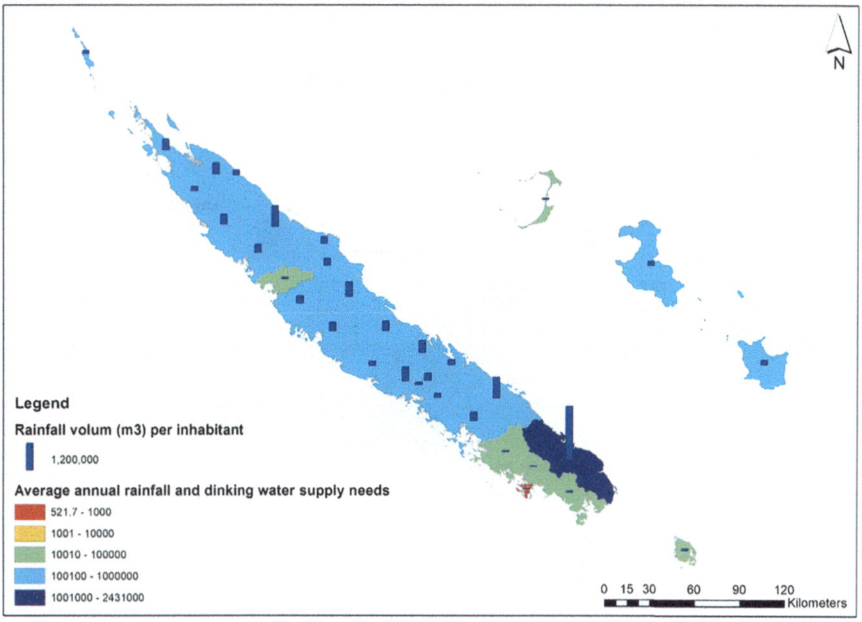

Fig. 3.2 Geographical disconnection between rainfall inputs and domestic water needs (DIMENC-SGNC 2023, based on data from ISEE 2019 and DAVAR)

Even though the population has doubled in 35 years and GDP has increased by 1.75 in 15 years (up to may 2024), demographic and economic pressures on the environment remain relatively weak. But these pressures are increasing, with water needs and consumption remaining for the most part still poorly known and authorised or estimated volumes are often not measured.

The map below (Fig. 3.3) shows the administrative water withdrawal authorisations by volume and category of use.

As shown in the previous figure, water collected in New Caledonia has three main uses: drinking water (80 million m^3/year), agriculture (70 million m^3/year), and industry (55 million m^3/year). The numbers in parentheses correspond to the volumes of water withdrawal administratively authorised, but these values must be increased to include non-regularised and "wild" samples. Agriculture and industry represent the two main economic uses for water in New Caledonia, and distribution according to uses differ for each of the watersheds and regions. Thus, the Greater Nouméa agglomeration mainly uses water for drinking water supply purposes, whereas the South-West region (Boulouparis, La Foa, Bourail) mainly uses water for agricultural purposes.

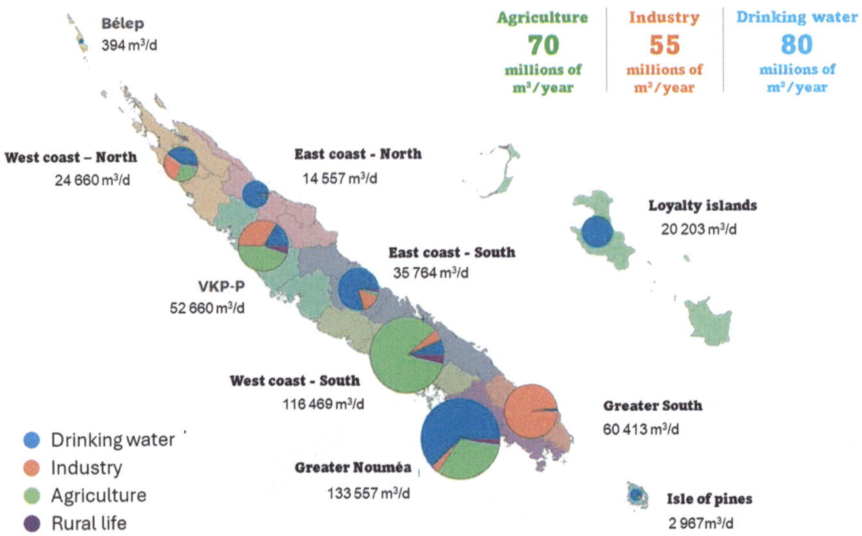

Fig. 3.3 Known and authorised volumes of water withdrawals/extractions, by regions and uses

3.2 Uses and Challenges

3.2.1 Domestic and Drinking Water to Satisfy Human Needs

While the majority of New Caledonia's inhabitants have access to a water with bacteriological qualities that meets international drinking thresholds/standards, difficulties exist in many rural and isolated areas of the country (see Chap. 4). Their extent is poorly known, because regular monitoring of the quality of distributed water is not organised in every village and tribe, and epidemiological surveillance does not include the monitoring of gastric disorders, which could serve as valuable health indicators.

As mentioned in Chap. 2, surface water (rivers or accompanying layers) are currently the main resources exploited in New Caledonia to meet domestic and economic needs. Extraction of groundwater by drilling mainly concerns islands with a lens of fresh water and the relatively arid coastal areas of the west coast (Fig. 3.4).

Surface withdrawals are particularly sensitive to sources of physical, chemical, parasitic, bacterial or virus contamination as well as drought, and watersheds are subject to various pressures which can have harmful effects for water quality consumption.

The watersheds which supply the drinking water catchments are generally not heavily urbanised or industrialised, so they appear to be less vulnerable to health risks associated with economic activities and domestic use. However, according to a study conducted on behalf of WWF in 2016, 90% of water protection perimeter (PPE) surfaces are in a degraded state, and half of them are even considered to be

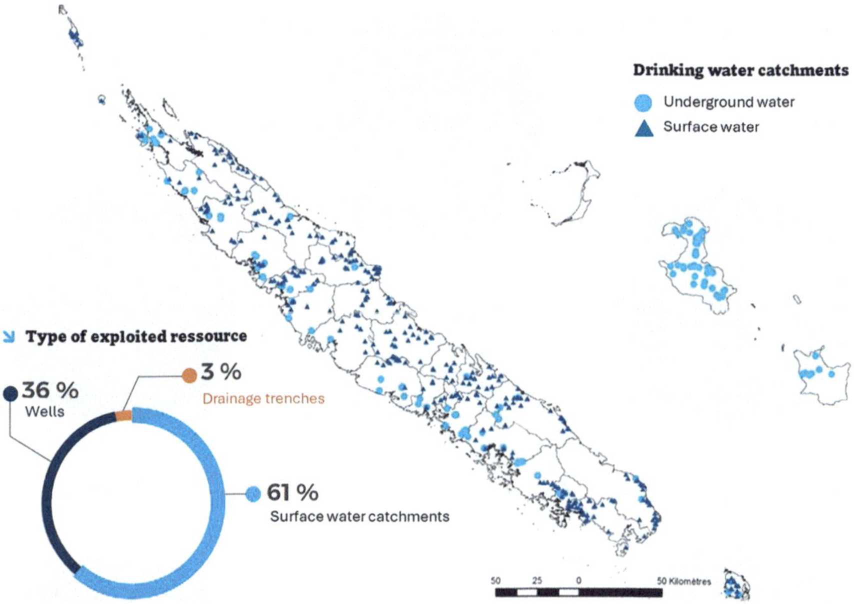

Fig. 3.4 Drinking water withdrawals in New Caledonia, with 64% of water catchments drawing on surface water and 36% on underground water

very degraded. This study considered criteria such as the condition of the forest cover, the sensitivity to erosion and some landscape and geology features. When the protection perimeter's forests and soil are damaged, the water quality degrades, increasing the risk of harm to public health.

Furthermore, in rural communities, the majority of the population does not pay for water. Rural municipalities estimate between 60% to 95% the percentage of their population which does not pay for the water. This is sometimes due to historical agreements with the tribes from which the water is sourced, sometimes due to a cultural perception that water is free, without an awareness of the cost of infrastructure, and sometimes due to technical and logistical challenges in reaching the town hall to pay the bill.

With the exception of the Greater Nouméa, municipalities resort to subsidies to balance their water and sanitation ancillary budget, which in principle is not permitted by the rules of the Code of Municipalities of New Caledonia and public accounting for municipalities with more than 3500 inhabitants. This subsidy can cover up to 80% of expenses when billing is not used, making the cost of water no longer borne by the consumer user, but by the taxpayer.

Non-payment of bills is one of the causes of over-consumption of water, and at the origin of the "vicious circle" of the drinking water service in New Caledonia, as shown in Fig. 3.5.

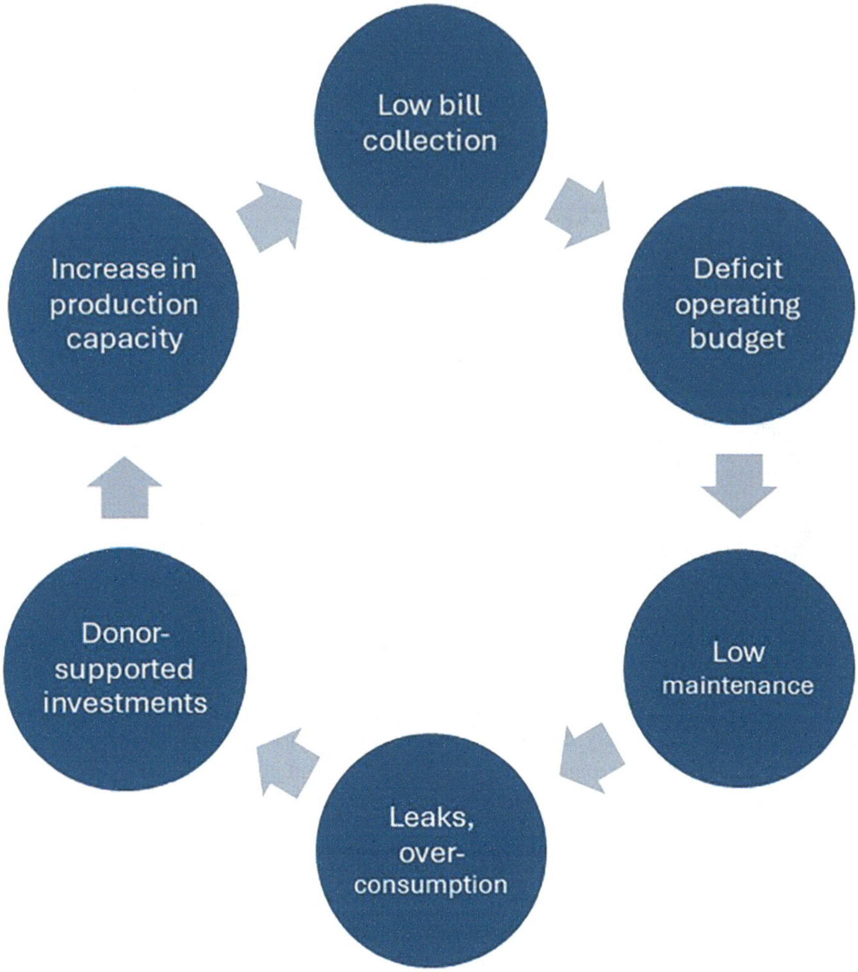

Fig. 3.5 The "vicious circle" of the service of drinking water in New Caledonia

3.2.2 Water, Agriculture and Food Security Rate

Ensuring the food security of its population is a major concern for any country. The insular nature of New Caledonia further emphasises this necessity, given its reliance on transportation and vulnerability to significant climatic events. Whenever feasible from a technical and economic standpoint, substituting imports with local production also addresses an economic development challenge by creating jobs and wealth, while simultaneously mitigating health and ecological risks.

The food security rate of the New Caledonian population through local production is an indicator of its supply safety. Excluding self-consumption, it was 28.4% in 2017 for fresh products (13.5% for processed products), with significant disparities across different sectors: over 70% for vegetables, approximately 55% for tubers, 45% for fruits, and only 33% for meat production. This discrepancy is partly due to the impact of recent droughts on cattle farming.

The country's informal economy, which is prevalent along the east coast and islands where residents primarily source their goods outside official channels, sometimes substantially inflates these coverage rates. It raises them to 80% for tubers over 70% for fruits, and 45% for meat (Bouard et al., 2023). However, there is still considerable scope for improving the food security of the people of New Caledonia.

This agricultural development challenge involves mastering and improving production methods, particularly through water resource management. It is critical for bolstering local production capacities and improving food coverage rates. Today, agriculture is the second user of water in New Caledonia, most of the time for crops that require a lot of water, like corn (subsided and used primarily to feed the livestock). Agricultors do not pay for the water that they pump directly from the rivers or underground wells, and overconsumption is pretty common. Water management with connected water metres and humidity sensors have started to be implemented in pilot farms since the beginning of the year 2020, but are far from being generalised throughout the country.

3.2.2.1 The Necessary Pursuit of Alignment Between Water Resources and Agricultural Needs

Figure 3.1 illustrates the current distribution of water volume withdrawals. The portions of New Caledonia's territory that are well-endowed with water and have potential storage spaces are not connected to areas of demand and usage. Subjected to climatic uncertainties, some watersheds experience frequent water deficits, resulting in significant reductions in river flow rates and even temporary drying up. This can lead to irreversible consequences for ecosystems.

The South Province has studied the equilibrium between needs and resources for a majority of its agricultural watersheds. These analyses have mapped the position of the saltwater wedge and developed a plan for hillside reservoirs as well as the economic development along the Ouenghi and Tontouta rivers. However, the irrigated agricultural areas associated with the Tamoa and Ouaménie hillside reservoirs, established in the 1990s, face significant operational management challenges.

Therefore, it is crucial to define a more efficient water withdrawal management approach to fully utilise existing and future infrastructure. In the North Province, the multi-purpose dam project in Pouembout is expected to cover domestic, environmental, and economic needs in the VKP area, particularly those related to agriculture, with the prospect of irrigating 4,300 hectares of farmland.

In the Loyalty Islands, the high sensitivity of the underground water table in Lifou and Maré (freshwater lens) to pollution risks precludes certain agricultural practices that would fail to ensure its preservation. Consequently, it strongly discourages the use of fertilisers or pesticides.

3.2.2.2 Water, a Critical But Often Limiting Production Factor in the Development of Agriculture in New Caledonia

Most commercial farms are located on the west coast and in the north of the main island, in areas highly vulnerable to climate threats. This observation correlates with land ownership maps, as the majority of private lands are situated in these parts of New Caledonia's territory.

The intensive cultivation period extends from May to September, during which there is a high demand for water in order to irrigate agricultural lands. Given that alluvial aquifers are limited in size, they serve as reservoirs with restricted capacity. Insufficient water availability exposes farmers to crop losses and reduced livestock performance. The low-water period (dry season) occurs at the end of the year (October to December).

On the west coast, which faces seasonal droughts, agricultural production—especially vegetables and fruits—tends to thrive in alluvial plains where topography and proximity to water are conducive to cultivation. However, this comes with the risk of exposure to flood damage. These agricultural systems remain highly vulnerable, and the economic compensation for the risks they face also impacts public funds.

Livestock watering represents the main water requirement for livestock farming. Pastures are not irrigated, with the possible exception of plots exclusively dedicated to hay production. Natural supplies would in any case be insufficient to consider intensifying fodder production through irrigation during low water periods and dry season.

In the context of large-scale crop systems in west coast regions like Tontouta, Boulouparis, Pouembout, and even Ouaco and Kaala-Gomen, water remains a limiting factor. Developing cultivation systems adapted to local conditions is crucial for effective management of agricultural water.

To enhance New Caledonia's food coverage rate:

- Agriculture must adapt to the available resources by influencing production systems (such as selecting suitable crop varieties and farming methods).
- Water should be harnessed for agriculture through storage or transfer methods, while also managing health and environmental risks.

The most fertile alluvial plains are also the most vulnerable to flooding. Mitigating this risk would reduce market disruptions caused by production losses but raises questions about land access and water resources.

3.2.3 Water and Mines

Several industries requiring substantial quantities of water exist in New Caledonia, including nickel mining and processing plants, breweries and golf courses.

The surfaces allocated to mining activities cover 15% of the Caledonian territory (2750 km^2 out of 18,350 km^2). 7.3% of these concession areas have been or are currently being exploited for mining, or they exhibit surface degradation related to recent or past mining activities (according to the SPOT 2005 study). In total, the areas of bare soil associated with mines or degraded due to mining activities cover approximately 200 km^2 (which is 1.2% of New Caledonia's total land area). Except for the Massif du Sud, these areas are situated along mountain ridges, potentially impacting multiple watersheds and the entire length of their watercourses.

Mining activities can have countless possible impacts. In many locations, during flood events, materials from old dumps and ravines cascade into riverbeds, sometimes obstructing water flow and compromising aquatic life due to excessive material influx.

Furthermore, in mining areas, the reduction in available water storage volume due to ongoing mining operations and the risk of mobilisation of naturally occurring toxic metals (such as environmental asbestos or arsenic) due to the spread of displaced materials or the risks of accidental pollution are significant potential threats to the underground resource. Environmental measures mandated by the mining code aim to address and remedy these risks.

Finally, downstream from these massifs, a significant portion of the recharge areas for New Caledonia's underground aquifers depends on the resources of the peridotite massif. These recharge areas are affected by mining development. For instance, consider the Tontouta catchment, a strategic resource for Greater Nouméa, where the aquifer located in the alluvium of the river at the base of the peridotite massif has 35% of its recharge area allocated to mining concessions (see Fig. 3.6).

While 10% of all inventoried groundwater intakes in 2018 are situated downstream from mining concessions, 33% of surface water intakes have mining concessions within their catchment areas. These waters are directly exposed to the suspension of fine particles, resulting from both natural erosion and mining activity during runoff events. These suspended materials become more noticeable when they are not properly managed on-site.

This situation raises questions about the long-term planning of resource use and the balancing of various uses. Several existing catchments within mining areas occasionally exceed metal limit concentrations allowed by New Caledonia's regulations, which are notably more tolerant for chromium and nickel than the World Health Organization standards. Thresholds can also be exceeded in areas without mining concessions or activity (as seen on Isle of Pines, for example). Depending on the circumstances, elevated concentrations of metallic elements in water can be attributed to human activity, natural geochemical background, or a combination of both (Cordier et al., 2017; see also Chap. 4).

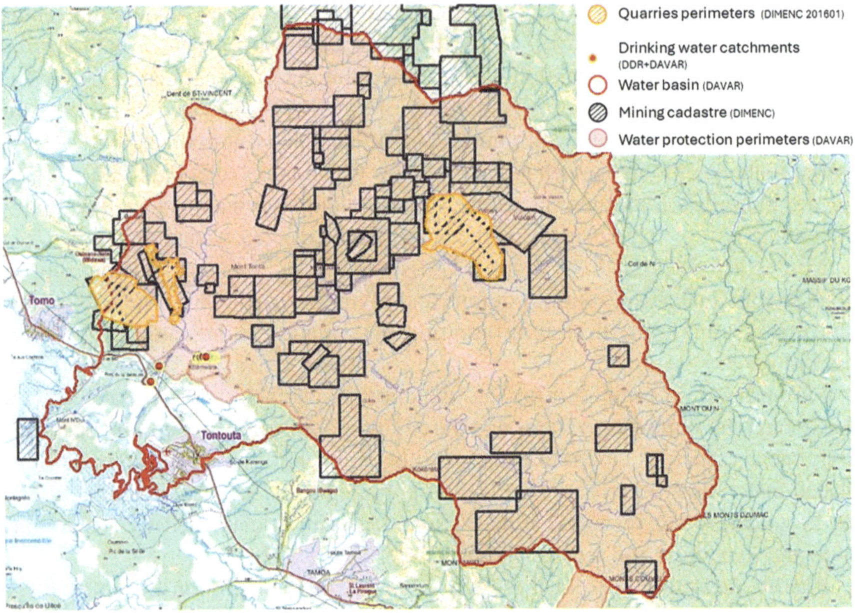

Fig. 3.6 Mining concessions, exploitations and water protection perimeter in Tontouta's watershed (source: DIMENC, DAVAR)

3.2.4 Other Economic Uses : Tourism and Hydroelectricity

Interestingly, New Caledonia's tourism industry and other economic activities have shown little interest in terrestrial waters and do not promote New Caledonian water resources as such, even though water remains an essential factor for production and economic development. Almost no tourism professional responded to the various consultations carried out during the development phase of the water policy in 2018/2019.

In some tourism hotspots in New Caledonia, water only becomes an issue when it begins to pose a problem: on the Isle of Pines, drinking water is starting to run out and led to a search for alternative sources via hydrogeological prospecting and desalination investments; in the Domaine de Déva tourist area, in Bourail, eutrophication phenomena have been suspected since the appearance of green algae in the lagoon.

In 2024, with an annual average production of 347 GWh (for 80 MW of installed power), hydroelectricity constitutes the largest source of production of renewable energy in New Caledonia, and answers between 10–20% of the country's energy needs depending on the hydrological inputs available. The hydraulic power constitutes a renewable energy particularly interesting to regulate electricity production, mainly obtained through coal and oil combustion.

Valorization of hydroelectric potential constitutes a strategic issue for the stability and security of the electricity network from the country. So, the *schema de la transition énergétique de Nouvelle-Calédonie* (STENC, public policy for the energy transition, voted by the congress of New Caledonia on June 23, 2016) aims to double hydroelectric power by 2030. However, if hydropower has a reputation for being a "clean" energy, it is not without impact on water resources and aquatic ecosystems. Indeed, the dams necessary for storage constitute almost impermeable obstacles to the sediment and biological flows of rivers. They thus significantly disrupt natural balances, with consequences that are often difficult to predict.

3.2.5 Preserving Water for Environmental Needs: Adapting the Minimum Biological Flow

With regard to environmental use, determining the minimum water requirement necessary for the survival of the ecosystem is challenging, and has only started to be taken into account in water withdrawal authorisations in the last couple of decades. A project, led by the French Biodiversity Office and the DAVAR Water Service, is currently underway, and aims to define a minimum biological flow for the watercourses of New Caledonia, in order to better adapt volume withdrawal authorisations to the available quantity of the resource.

The definition of the minimal biological flow is different for every river, and depends on geography, topography, average annual rainfall, species that are present in the water basin, urbanization, anthropogenic impacts. A multi-stake approach is needed, taking into account the ecological, hydrological, geomorphological, legal and economical aspects. Experts need to work together to find the right equilibrium between ecosystem preservation and human needs for water.

3.3 Tensions and Threats

3.3.1 Shortages and Flooding Exacerbated by Climate Change

Precipitation is very unevenly distributed over time and space in New Caledonia. Climate fluctuations expose New Caledonia to episodes of drought and flooding.

From water deficit…

The high variability of rainfall inputs exposes New Caledonia to severe low-water periods (droughts) and recurrent risks of imbalance between resources and usage. These periods of "drought" generally occur during the last quarter of the year, when river flow rates are five to ten times lower than during the first quarter.

Due to the limited size of the watersheds, combined with the low storage capacity of underground reservoirs (such as aquifers), many watercourses experience complete drying up (or "assec") after 90 days without rain. This strong dependence on water resources and aquatic environments is closely tied to meteorological conditions.

This quantitative vulnerability is even more significant in areas where usage is concentrated, particularly along the west coast, where cumulative rainfall is least abundant. Water withdrawals, especially for agricultural purposes, as well as domestic and industrial needs, peak during low-water periods, resulting in significant management and planning challenges.

… To water excess

While the last quarter of the year is characterised by a significant risk of water deficit, the first quarter, and more broadly, the cyclone season, expose New Caledonia to a particularly high risk of excessive water and flooding. Precipitation exceeding 100 mm in 24 h is commonly observed annually, with records typically occurring during cyclonic episodes (such as the 1988 Cyclone "Anne", which brought over 900 mm of rain in 24 h). These extreme intensities, coupled with cyclonic phenomena (wind and sea level rise), result in peak flood flow rates close to global records (easily surpassing 20 $m^3/s/km^2$), propagating from ridges to lagoons within a matter of hours. These are therefore "flash floods", generating flood levels frequently exceeding more than one meter in the alluvial plains.

This hazard has significant implications. Over 20,000 homes and a large majority of cultivated areas are directly affected by the risk of flooding. Depending on the steepness of the terrain, residential zones may also be exposed to the risk of land movements, related to slope saturation during heavy rains, especially in the lateritic soils of the mining sector.

Given this climatic constraint, policies of prevention and development are essential to safeguard civil security and economic interests (Dutheil et al., 2020)

The first climate change forecasts are predicting an increase in temperature, increasing the droughts risks, with many impacts for human health and agriculture. For New Caledonia, air temperature is forecasted to increase by +2 °C in 2040. Sea temperature will also increase, leading to an increase in the number of tropical cyclones. And experts predict that the average rainfall from cyclones will strongly rise, which will lead to more floods.[1]

3.3.2 *Growing Environmental Pressures*

3.3.2.1 Fires and Invasive Species

On the scale of the island, the environmental observatory (ŒIL) estimates that degraded habitats such as scrubland and savannah increased by 2.6–52%. Bare soil resulting from 130 years of mining activity represents 20,000 ha, or 1.2% of the territory and the silting in rivers and creeks has been estimated to reach 600 km

[1] Secretariat of the Pacific Regional Environment Programme, Pacific Islands Climate Change Monitor, 2021, completed by studies by l'Institut de la Recherche et du Développement NC and Météo-France NC for the Revue Juridique, Politique et Économique de Nouvelle-Calédonie n°41, January 2023.

(PEP, 2019). The burnt areas are estimated between 10,000 and 30,000 ha annually and the origin is 99% human (OEIL, 2023).

New Caledonia's ecosystems and biodiversity are also struggling with many invasive species, in particular the deer Rusa, imported as a gift from Indonesia in the 1960's, and wild pigs. The deer population, difficult to quantitify, is estimated to exceed 300,000 animals, despite ongoing collection and regulation efforts. This number surpasses the last (human) census of 2019 which announced New Caledonia's population to be around 271,000 inhabitants.

Deers and wild pigs have no local predators, and they have been demolishing the tropical forests undergrowth by eating roots and young shoots. The ground degradation is not visible from the satellite images, so the extent to the damage has been taken into account only very recently, from participative field trips organised with the local populations.

The forest undergrowth's disappearance created land erosion under the big trees: the soils are bare when you hike in many of the forests, especially on the west coast of the main island. This soil is then easily washed away during heavy rains, eroding more and more of the first layers of earth, making the forests' regeneration more and more difficult. This accelerates the forests' disappearance and reduces the roots and soil power of water conservation, shrinking the "natural sponge" role of the tropical forests.

3.3.2.2 Pressure on Water Quality Linked to Economic Activities and Absence of Sanitation Policy

We previously discussed the quantitative threats on the water resource due to economic activities: mining, industry, agriculture, hydroelectricity, tourism, etc. These human activities also directly impact water quality through their discharges. We must also consider the discharges of wastewater from both collective and non-collective sewage systems, which we have not yet mentioned.

The main issue related to water quality in New Caledonia is turbidity. Suspended solids, which originate from the leaching of eroded forest soils due to invasive species, from fires and mines, and "red waters," as referred to by local populations, cause high turbidity in surface waters during each rainfall event. Water supply intakes regularly fill up with solid materials, reducing the available quantity of water for extraction and increasing filtration needs prior to distribution.

In most rural communes, municipalities responsible for distributing drinking water are unable to supply their populations for several days during each rainfall event and must regularly carry out costly dredging operations. Tribes and local populations are accustomed to this situation, which should normally only occur during major crises. It should be noted that this is combined with very high consumption rates that do not allow installed reservoirs to act as a buffer between water supply and demand.

"Red waters" are most frequently present in the watersheds of mining communes of the East Coast, including Yaté, Thio, Canala, Kouaoua, Houaïlou, and

Ponérihouen, as well as Koné, Voh, Koumac, Poum, Poya, and the Tontouta River on the West Coast. Indeed, the stripping of watershed ridges for nickel extraction has completely eliminated forests in exploited areas, and water management on mines has yet to contain all fine or coarse particles during rainfall events. Furthermore, New Caledonia regularly experiences landslides and the destruction of sedimentation basins, which create gullies and ravines in mountain ranges and result in impressive amounts of solid materials (stones, rocks) entering downstream watercourses. Tomûrû River in Thio is just one example of a watercourse that has become a river of pebbles.

These "red waters" degrade the drinking water quality, leading to numerous episodes of diarrhea or gastroenteritis. Unfortunately, these diseases are not recorded and monitored by medical services, making it impossible to clearly identify and quantify water quality-related epidemics on a large scale with an accurate historical and geographical analysis.

Many industries discharge their wastewater into rivers, and the physico-chemical states of these discharges are not regularly monitored. In fact, only the regulations related to Classified Installations for the Environment[2] currently require industries to conduct regular analyses of their discharges, but obtaining data from actors is actually difficult. This difficulty is also present for discharges from wastewater treatment plants for collective sewage. It is therefore almost impossible to know if diffuse pollution exists in an industrial discharge, and alerts are only given when organoleptic parameters (odor, water color) become unpleasant for neighboring populations, or when they witness mass fish deaths. The government services of New Caledonia receive an average of ten alerts of this type per year.

Another type of diffuse pollution that New Caledonia must address is that of molecules from pesticides, fertilizers, and other phytosanitary products used in agriculture, which then spread into rivers and groundwater. The Chamber of Agriculture and agricultural professional organizations are leading awareness and training efforts on the use of these environmentally harmful products to reduce their use at the source and thus reduce the potential for diffuse pollution in soils and the surrounding environment of cultivated plots.

Apart from the regulations related to Classified Installations for the Environment, there are no regulations related to sanitation, whether non-collective (septic tanks) or collective (wastewater treatment plants). The installation of a septic tank is not mandatory at the country level, only strongly recommended in the New Caledonia Construction Reference, and some municipalities have taken regulatory measures to require their installation on their territory. Similarly, connection to collective

[2] Installations Classées pour la Protection de l'Environnement (ICPE): réglementation et contrôle des ICPE; Réglementations relatives à la gestion des déchets; Code de l'Environnement de la Province Sud; Code de l'environnement de la Province Nord et Code de l'environnement de la Province des îles Loyauté. (Regulation and control of Classified Installations for Environmental Protection; Waste management regulations; Environmental Code of the South Province; Environmental Code of the North Province; and Environmental Code of the province of the Loyalty Islands.)

sewage is not mandatory and is often not carried out by property owners, resulting in wastewater treatment plants operating at less than capacity. Studies and consultation workshops were conducted on this topic in 2009 with the "OPANC: Public Office of Sanitation in New Caledonia" project. Unfortunately, like many public policies in New Caledonia, it remained at the prefiguration stage due to a lack of strong and assumed political support. However, with the adoption of the Shared Water Policy and the creation of the working group "Strategic Objective 4: Move towards zero discharge of untreated water," work around pollution and waterwater may be able to resume with the support of the DASS-NC, in partnership with the French Office of Biodiversity.

3.3.3 Emergence of Social Tensions

Even if water is not an important source of social tensions in New Caledonia, local tensions exist and have emerged recently. The map below sums up some of these local tensions (Fig. 3.7).

Local conflicts emerged around pollution, and in particular pollution resulting from mining activities, around the payment of water bills and the over-consumption of drinking water and regarding "sacred" water (as certain springs or water holes are prohibited or taboo[3]). Like land and natural resources, water is an element of common heritage and cultural identity. Any project aimed at installing water networks is subject to the prior consent of the chiefdom[3] and the clans and is granted on the basis of the results of an environmental and socio-cultural impact analysis (Customary Senate, 2014). As within other indigenous populations of the Pacific, South America or Canada, water and its associated infrastructures could become a source of grievances and conflict in New Caledonia.

3.4 Conclusion

Water resources in New Caledonia are thus extremely unevenly distributed, both in terms of time and space, and generally in stark contrast to water needs. Water usages are very often in opposition with available resources: "We need water the most where it is scarce, precisely when it is least available!". This situation significantly restricts usage, especially considering its importance in shaping development and planning policies.

[3]The notions of "sacred" and "taboo" are polysemous and dynamic (Leblic, 2005). The conditions of the restrictions and access to the sites differ from one individual or clan to another. Non-compliance with these restrictions results in sanctions which vary according to the sites and the people who violate the restrictions.

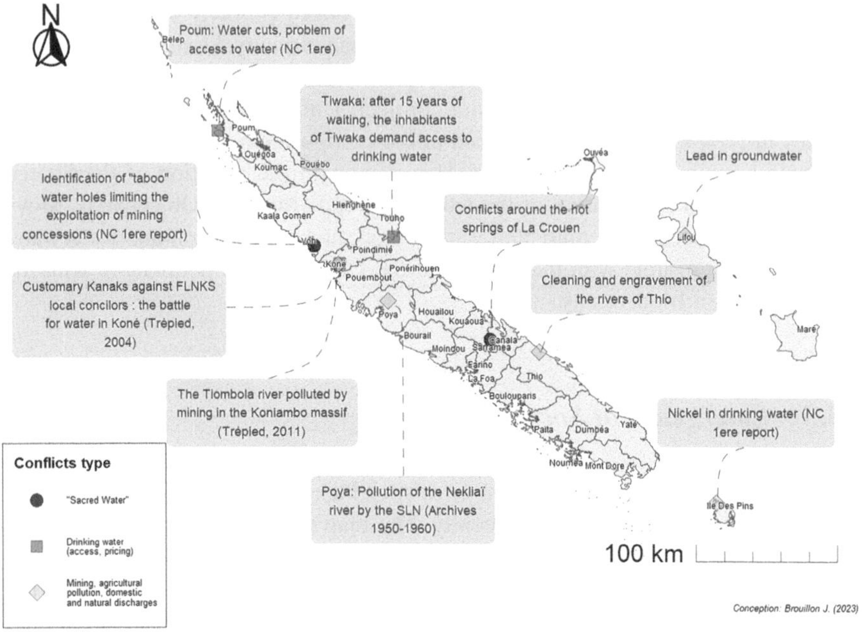

Fig. 3.7 Emerging social tensions

Overcoming the constraint of water scarcity involves considering the construction of artificial reservoirs capable of retaining excess water during the wet season for use during the dry season. This is why there is a strong need for structured development to mitigate constraints on the three water usages: domestic, economic, and environmental.

Water sharing requires defining cooperation modalities among stakeholders. Within the same watershed, it involves respecting hydrographic patterns and establishing an equitable distribution of usage, ensuring a fair allocation of resources. However, this cooperation can also extend across different watersheds, where the strengths of one area can compensate for the weaknesses of another. For this to succeed, relationships of solidarity must remain balanced. Achieving this ambition depends on the ability of decision-makers to establish governance commensurate with the challenge.

This ambition has been incorporated into a draft territorial law (Loi de Pays) related to the public domain of water. Written collaboratively by more than 10 institutions and directorates since December 2020, this draft territorial law explicitly allows for the establishment of local water councils at watershed or municipality level, paving the way for integrated water resource management. Unfortunately, as of March 2024, this draft law has not yet been adopted by the Congress of New Caledonia, due to a lack of genuine political support and consensus.

References

Bouard, S., Apithy, L., Guyard, S., Passouant, M., Sourisseau, J. M., & Bélières, J. F. (2023). Informal and non-market operations in activity systems: representative survey of Kanak households in the indigenous villages of New Caledonia. *Open Data Journal for Agricultural Research, 9*, 1–9. https://doi.org/10.18174/odjar.v9i0.16396

Cordier, S., Barguil, Y., Dominique, Y., St-Jean, A., Mengant, S., Le Bot, B., & Ayotte, P. (2017). *Niveaux d'imprégnation et déterminants de l'exposition humaine aux métaux en Nouvelle-Calédonie. Programme "METEXPO"*. CNRT Nickel & Son Environment. 240 pages.

Dutheil, C., Lengaigne, M., Bador, M., et al. (2020). Impact of projected sea surface temperature biases on tropical cyclones projections in the South Pacific. *Scientific Reports, 10*, 4838. https://doi.org/10.1038/s41598-020-61570-6

ISEE (2023). *Institut de la statistique et des études économiques de la Nouvelle Calédonie*. Retrieved from https://www.isee.nc/

Leblic, I. (2005, 2008). Pays « surnature » et sites « sacrés » paicî à Ponérihouen (Nouvelle-Calédonie). *Journal de la société des océanistes, (120–121)*, 95–111. https://doi.org/10.4000/jso.299, https://doi.org/10.4000/jso.410

OEIL (2023). *Observatoire de l'Environnement de Nouvelle Calédonie*. Retrieved from https://oeil.nc

PEP (2019). *Schéma d'Orientation de la Politique de l'Eau Partagée en Nouvelle Calédonie*. Retrieved from https://eau.nc/sites/default/files/documents/pep_-_schema_dorientation_pour_une_politique_de_leau_partagee_de_la_nouvelle-caledonie.pdf

Further Reading

Fallon, G., Wotling, G., Winchester, P., et al. (2019). Schéma pour une politique de l'eau partagée de la Nouvelle-Calédonie, gouvernement de la Nouvelle-Calédonie, mars 2019. Figures 3.1, 3.3, 3.4 and 3.6 are extracted from PEP-NC and translated by authors.

Gouvernement de la Nouvelle-Calédonie, Sénat Coutumier, Charte du peuple kanak, délibération n°06-2014/SC du 15 juillet 2014 constatant la proclamation et portant adoption de la charte du peuple kanak fixant le socle commun des valeurs et principes fondamentaux de la civilisation kanak. https://juridoc.gouv.nc/juridoc/jdwebe.nsf/joncentry?openpage&ap=2014&page=6815

Chapter 4
Water Quality: Trends and Challenges

Karine Saumë and Charlotte Duval

Abstract Today, New Caledonia's drinking water supply comes from the freshwater lens on the Loyalty Islands, from surface water on the east coast, and from alluvial groundwater at the base of the massifs, and especially peridotites, on the west coast. On Grande Terre (the main island), drinking water comes mainly from surface water, which is the most vulnerable to pollution, floods and droughts. Yet, of the approximately 500 catchments for drinking water, 31% distribute untreated water and only 54% of catchments have a protection perimeter (New Caledonia, 2019). Moreover, the current standards concerning the quality of water intended for human consumption were set by amended order no. 79-153 of 3 April 1979 which retains, for 15 referenced parameters, limits that are mostly higher than those defined by the metropolitan regulations and the World Health Organization (WHO). In this context, this chapter presents the data available on water quality in New Caledonia. It describes the complex dynamics between natural factors (geology, hydrology) and human factor (mining, agriculture, urban space). It underlines the diverse range of challenges related to water quality especially the natural high level of heavy metals. The presence of metals (nickel, cobalt, chromium, iron, manganese) in sometimes high concentrations in the Caledonian natural environment (soils and subsoils) constitutes a geochemical "background noise" which can also affect drinking water and food. There are several challenges for water policy in terms of water quality, including strengthening the protection of reservoirs and rivers, and establishing a more comprehensive and dense monitoring network. Water treatment or replacement of non-compliant water is a major technical and financial challenge.

Keywords Drinking water · Water quality · Water sanitation

K. Saumë (✉) · C. Duval
DASS, Directorate of Sanitary and Social Affairs, Government of New Caledonia, Nouméa, New Caledonia
e-mail: karine.saume@gouv.nc; charlotte.duval@gouv.nc

C. Lejars, S. Bouard (eds.), *Water Policy in New Caledonia*, Global Issues in Water Policy 32, https://doi.org/10.1007/978-3-031-88196-1_4

4.1 Introduction

The quality of drinking water is a public health issue in many parts of the world, including New Caledonia. Difficulties in accessing drinking water and basic hygiene conditions remain major factors in water-borne diseases in New Caledonia which has experienced epidemics of hepatitis A, annual outbreaks of gastroenteritis and cases of amoebiasis, among others. This demonstrates the clear need to put water and health back at the centre of the debate.

While the concept of drinking water seemed to be taken for granted by everyone, the very first edition of the water/health barometer carried out by New Caledonia in 2020 with the financial support of the 11th European Development Fund (EDF), still shows major disparities in the fundamental understanding of potability and associated health risks.

Given these observations, it is clear that the issues surrounding drinking water are multi-sectoral and need to be addressed across the board. It was against this backdrop that action 60 of the DO KAMO New Caledonian Health Plan[1] was included, with the aim of integrating drinking water quality, control and health risk management into the future Shared Water Policy (PEP), thereby combining environmental approaches with health objectives and illustrating the desire for congruence between New Caledonian public policies.

4.2 Issues and Challenges

4.2.1 The General Framework for Drinking Water in New Caledonia

In New Caledonia, even though drinking water distribution is defined by the Municipalities Code as being the responsibility of the mayors, there are currently no framework regulations in terms of the safety of drinking water, setting out the obligations of suppliers, the management, monitoring or quality control procedures for the water produced and distributed. The only existing New Caledonian regulations on drinking water currently concern the potability standards introduced by the amended Order No. 79-153 of 3 April 1979. Outdated and obsolete, this text requires major updating to bring it in line with international health recommendations, while taking into account New Caledonia's unique situation. For the time being, therefore, appendix 1 of the French decree of 11 January 2007 serves as a reference, although it is not legally applicable due to the status of New Caledonia.

As a result of a fragmented and obsolete regulatory framework, the quality of tap water remains very heterogeneous in New Caledonia. Today, the main disparities

[1] A French version is available online: https://gouv.nc/dossiers/le-plan-de-sante-caledonian-do-kamo. See also Faberon (2017).

Table 4.1 A summary of the organisation of the drinking water supply in New Caledonia, based on a plot-by-plot inventory drawn up using the latest WSP data and interpretation of the DASS-NC database, health situation 2018

AEP uses	New Caledonia (33)	South Province (14)	North Province (16)	Islands Province (3)
Number of resources	444	178	191	75
Including catchments	238 (53.6%)	91 (51.1%)	147 (77%)	0
Including boreholes	185 (41.7%)	71 (39.9%)	39 (20.4%)	75 (100%)
Including drainage ditches	21 (4.7%)	16 (9%)	5 (2.6%)	0
Unit of Distribution (UD) number	266	82	135	49
With bacteriological treatment	165 (62%)	69 (84.1%)	63 (46.7%)	33 (67.3%)
With physico-chemical treatment	47 (17.7%)	12 (14.6%)	32 (23.7%)	3 (6.1%)
No treatment	81 (30.4%)	12 (14.6%)	53 (39.2%)	16 (32.6%)
Population concerned (population general census 2014, ISEE)	268,767	202,789	47,681	18,297

concern the treatments applied to the water (see Table 4.1). Sometimes unsuitable or even non-existent, these treatment faults expose consumers, particularly the most vulnerable, to proven health risks. This is compounded by an insufficient number of water quality checks, making it extremely difficult to identify and deal with the risk.

4.2.2 *Vulnerable Water Resources*

In New Caledonia, most of the resources used to produce drinking water are surface resources (53.6% of the country's total resources). Because of their configuration, these surface resources are much more exposed to external sources of contamination (particularly faecal pollution) and to meteorological events (drought, heavy rain, etc.), thus generating major variations in water quality, as illustrated in the graphs in Fig 4.1.

New Caledonia's particular geology also exposes its water resources to other quality problems, this time linked to the natural presence of heavy metals, particularly nickel and chromium (see Fig. 4.2).

The METEXPO study conducted between 2016 and 2018 showed that inhabitants of New Caledonia have a high level of exposure to metals and highlighted the decisive role of local geology in the contamination of all environments, thus contributing to increasing the toxic load of metals in the body, particularly nickel, by inhalation (dust) and orally (water, food, dust) (Cordier et al., 2017).

Even if the METEXPO study is not conclusive on the sources of exposure responsible for the increased levels found in populations, given that water resources that are sometimes extremely concentrated in metals, it nevertheless supports the observation that current New Caledonian standards (cf. Table 4.2) are seemingly inadequate to ensure the health safety of the water distributed. The challenge for

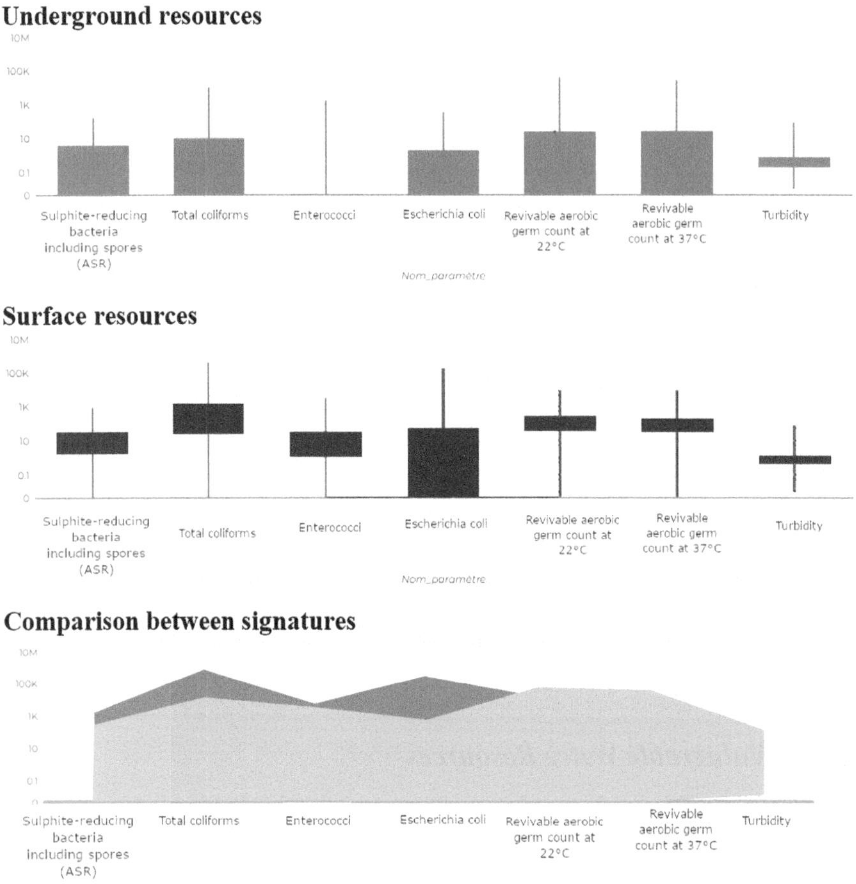

Fig. 4.1 Comparison of quality variations between underground and surface resources

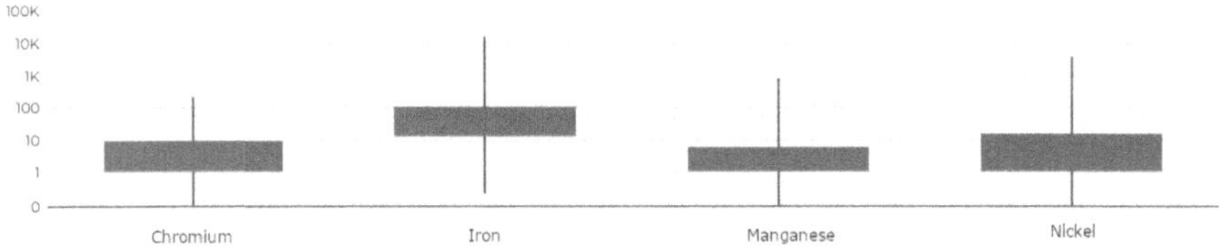

Fig. 4.2 Dispersion of specific metals in raw water resources used for drinking water production in New Caledonia

New Caledonia in the future will therefore be to define raw water and drinking water quality standards adapted to the specific context of the New Caledonian territory. This will involve making strategic choices in terms of water resource quality on a more localised scale, especially in order to minimise operating costs associated

Table 4.2 comparison of potability standards for metals in drinking water

Parameters	New-Caledonia Concentration en µg/L	Water safety standards changed in France in December 2022: selenium: 20 ug/L	WHO
Lead	100	10	10
Selenium	50	10	40
Fluorides	1000	1500	1500
Arsenic	50	10	10
Nickel	100	20	70
Chromium	50	50	50

with treatment. In any case, this is what New Caledonia has committed to through the Shared Water Policy.

4.2.3 Degradation Factors (Fires, Invasive Alien Species, Mining Activities)

In addition to the natural vulnerability of water resources, fires and invasive exotic species are major factors in the deterioration of these areas.

Fires, which are mainly of human origin (cultivation practices, hunting practices, inattention or malicious acts), destroy vegetation and expose soils to erosion, thereby compromising the recharge of natural water reservoirs and affecting their quality. Fires also have repercussions on the mobility and availability of certain trace metals, as confirmed by recent research showing that fires in ultrabasic catchments accelerate the dynamics of certain trace metals, thereby affecting the various uses of water, particularly those dedicated to the production of drinking water (Thery et al., 2023).

Invasive species, particularly deer and wild pigs, are also damaging catchment areas (browsing that hinders regeneration, debarking, trampling), making soils more susceptible to erosion and considerably disrupting the recharge of natural water reservoirs. The use of physical barriers to protect crops in lowland areas has driven these invasive species further into the mountain range, thereby intensifying their destructive impacts on water resources.

Mining also puts pressure on water resources. The surface area allocated to mining covers 15% of New Caledonia. There are many potential impacts. While water withdrawals for mining are limited, mining nevertheless diminishes water storage capacity and increases the risk of mobilising naturally occurring metals, potentially endangering water resources.

4.2.4 Unsafe Uses and Unequal Access to Drinking Water

In 2022, the DASS-NC database contained 196,176 test results collected between 1997 and 2018 in the 33 communes of New Caledonia. Due to the absence of framework regulations requiring, in particular, monitoring of the quality of the water distributed, the data sets for each commune are currently very fragmented and do not provide an accurate and up-to-date picture of the quality of water distributed from a spatial and temporal perspective.

Although patchy, the data collected by DASS-NC nonetheless demonstrate the impact of the lack of treatment infrastructure on the quality of the water distributed (see Figs. 4.3 and 4.4) and the proven exposure of populations to greater or lesser health risks depending on their locality, thereby demonstrating the inequality of access to drinking water in New Caledonia.

Following the validation of the PEP by the Congress of New Caledonia, the issues related to the quantity of water available to meet basic hydration and hygiene needs were also included in New Caledonia's roadmap. Against this backdrop, an initial diagnostic study has been launched in 2019 to quantify the "lack of water" health risk for New Caledonian households.

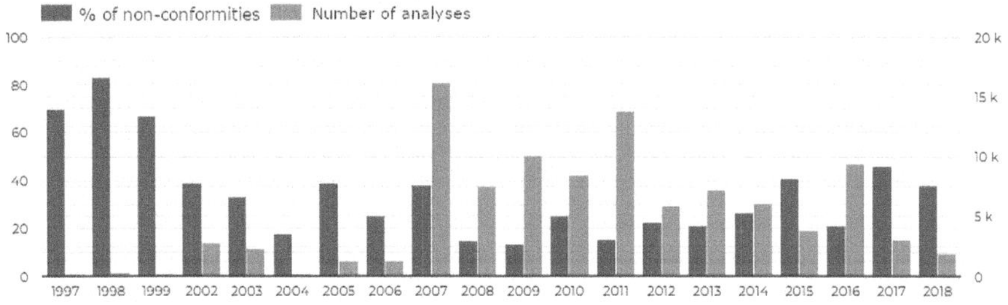

Fig. 4.3 Number of analyses collected and proportion of associated non-conformities referenced in the DASS-NC database between 1997 and 2018

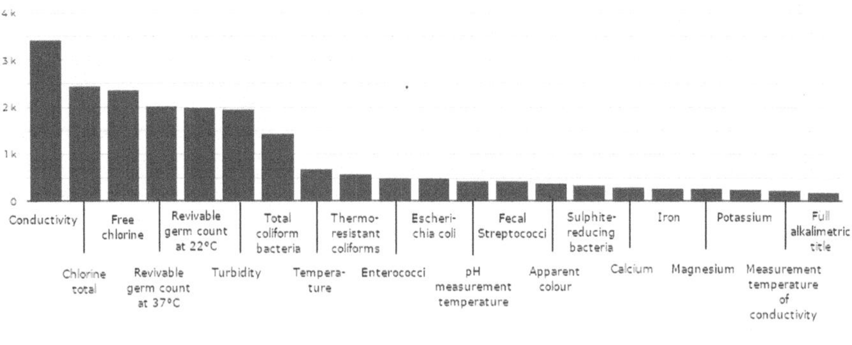

Fig. 4.4 Number of non-compliant quality analyses referenced in the DASS-NC database between 1997 and 2018

4.3 Existing Measures and Initiatives to Meet These Challenges

4.3.1 Measures to Protect Water Resources Need to Be Strengthened

Of all the resources used to produce drinking water in New Caledonia, 54% have a water protection perimeter and 39% are in the process of establishing one (source DAVAR, 2018). However, these regulatory tools, initially defined by New Caledonian deliberation No. 105 of 9 August 1968 and designed to preserve the quality of the resource, still have limitations when faced with the large number of abstraction works, particularly in terms of controls and policing.

According to a study carried out for WWF in 2016, 90% of water protection areas (WPA) are in a degraded state (half of them very degraded), based on criteria that take into account the state of the forest cover, susceptibility to erosion and landscape features (see Fig. 4.5). As soon as the protective cover is damaged, the risk of water quality deterioration increases, and with it the risk of harm to public health.

Fig. 4.5 Photographs of a catchment on Île des Pins after bushfires

4.3.2 *Insufficient Water Treatment Infrastructures to Cope with Pressures*

Bushfires, invasive species, urbanisation, erosion and human activities are all increasing pressures associated with water and generating numerous tensions on the resource. Global warming is intensifying these pressures by altering the chemical and biological balance of the environment, which encourages the proliferation of germs and bacteria that are potentially dangerous not only to ecosystems but also to human health.

Faced with these constant pressures, the risk is very real. According to the latest data collected by the DASS-NC, the water distribution networks in New Caledonia do not have sufficient treatment infrastructure to ensure that water distributed to consumers is safe for their health. Of the 266 water distribution units in the country:

- only 62% have disinfection treatment,
- 17.3% have physical treatment, mainly to reduce turbidity,
- 30.4% have no treatment at all and distribute raw water.

4.4 Support for Local Authorities in Controlling Health Risks

4.4.1 *Water Safety Plans (WSP)*

Between 2008 and 2018, New Caledonia provided technical and financial support to assist local authorities in managing the health risks associated with the water they supply. This was achieved by developing the Water Safety Plan (WSP) tool.

The WSP tool, developed by New Zealand, has been recognised by the World Health Organization (WHO) for its effectiveness. Indeed, the WHO Guidelines for Drinking Water Quality (2004), define the implementation of a water safety management plan as the most effective approach for ensuring a continuous supply of drinking water.

In the absence of a regulatory framework defining the health standards applicable to drinking water in New Caledonia, the objective of introducing the WSPs was to raise awareness among local authorities of the health risks associated with water in the network and to engage them in a comprehensive and continuous improvement process. In order to develop the WSP tool in New Caledonia, the DASS-NC teams spent several weeks training in New Zealand, as well as several months integrating the WSP specificities into a Caledonian version of the WSP tool. After several months of work, the inaugural Caledonian WSP was produced in 2008 for the municipality of La Foa.

While the conventional method of managing water quality focused mainly on monitoring water as a "finished product," the WSP tool promotes a different approach based on preventive risk management. Based on "quality assurance" type procedures, the WSP enables the identification and anticipation of critical points in the drinking water supply network at an early stage, and the planning of improvements to be implemented by the municipality to ensure the quality of the water distributed to consumers.

The development of a WSP is a process that occurs in a number of stages, as illustrated in Fig. 4.6 Each WSP commences with the formation of a multidisciplinary team, which may include elected representatives, water engineers, private service providers, health educators or doctors from dispensaries, emergency services, local authorities, and so forth. These individuals come from diverse backgrounds and contribute their expertise in technical, financial, social, and economic aspects necessary for the development and successful implementation of the tool. The composition of the WSP team is of paramount importance for the success of the project. It is the responsibility of the WSP team to ensure the smooth running and correct implementation of the project.

The development of the PSSE program (WSP) in New Caledonia has taken 10 years. At the end of 2018, all municipalities had their own tool. Combined, all the measures proposed in the improvement programs of the 33 WSPs totalled over 6 billion CFP francs.

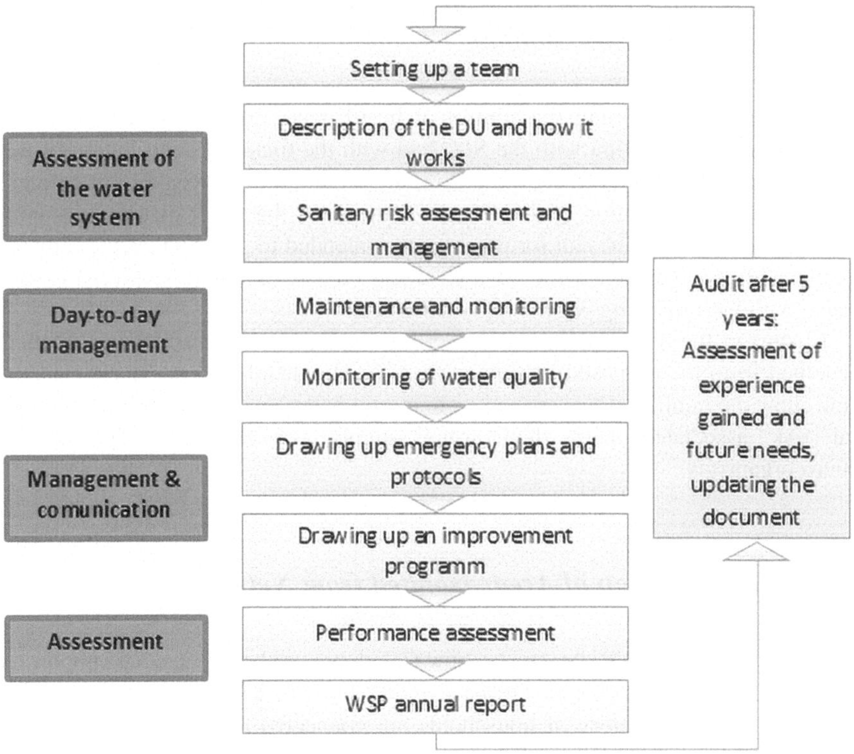

Fig. 4.6 Stages of the WSP tool

In New Caledonia, the preparation of the WSPs was part of a participatory and voluntary approach on the part of the municipalities. The idea is to provide support and increase their awareness of drinking water issues prior to the implementation of regulations. Despite the efforts of the various stakeholders, more than 15 years after the first WSPs were drawn up, the results remain mixed. The price of improvement programs as well as changes to teams within municipalities seem to have limited the implementation of recommendations and the follow-up of WSP actions. The absence of regulation and enforceable measures also seem to have been factors behind the inaction, or at least seem to have contributed to maintaining the status quo without any real incentive for action.

A few years later, in 2021, the tool was developed in metropolitan France as the Water Safety Management Plan (PGSSE), in addition to the regulations already in force in the metropolitan area. In the Pacific, it was from 2019, with the help of the 11th European Development Fund (PROTEGE program), that the WSP tool was extended to French Polynesia and Wallis and Futuna. A week of workshops took place in March 2024 with the 3 territories to share feedback on the tool, which has been finalised for some and in the early stages for others.

4.4.2 A Call for Projects for the Installation of Disinfection Stations

With a view to supporting the control of health risks, in 2020 New Caledonia launched, in collaboration with the SPC and with the financial contribution of the 11th EDF and the French Office of biodiversity, a call for projects aimed at municipalities for the installation of disinfection stations on the water distribution units deemed most at risk. The call for projects was intended to generate interest among municipalities and engage them into a voluntary process to reduce health risks associated with their drinking water supply networks.

12 sites in the municipalities of Ponérihouen, Canala, Thio and Kaala-Gomen benefited from the proposed funding (Fig. 4.7). The disinfection devices installed now allow the municipalities concerned to ensure better control of the bacteriological risk associated with their water supply by eliminating pathogenic micro-organisms.

4.4.3 Identification of Areas Isolated from Networks

In order to meet the PEP objectives, New Caledonia, with the financial support of the 11th EDF, also initiated the "isolated areas" study in 2019 intended to draw up a comprehensive inventory of households not connected to the communal water network and lacking a sustainable water resource. Intended mainly to quantify the

Fig. 4.7 Status of works on some sites to be equipped as part of the call for projects launched by the SPC and the DASS-NC (photo credit: Authors)

health risk of "lack of water" in New Caledonian households, the "isolated areas" study led by the DASS-NC also had a second objective, to develop a catalog of alternative water supply solutions for the municipal network, tailored to the identified situations (see Fig. 4.8) and ensuring safety from a health perspective.

The intermediate results of this study (cf. Fig. 4.9), which today concerns 26 of the 33 Caledonian communes (i.e. 30% of New Caledonian households), have made it possible to highlight, beyond the geolocation of the sites concerned, the following elements:

- 6.1% of households are not connected to the municipal distribution network,
- 0.9% of all surveyed households have no individual or collective water access point.

By way of comparison, in New Zealand, 10% of the population is not connected to communal networks.

The initial results of the study also highlight:

- The good knowledge that municipalities have about the water supply of households located on their territory, whether or not there are up-to-date planning documents. It should be noted that of the 26 municipalities encountered, 12 declared that they did not have a master urban planning plan (PUD), i.e. 46%. Of these 23 communes, 63% have a drinking water supply master plan (SDAEP) that is more than 5 years old or non-existent. This means that spatial planning in these municipalities is carried out without reference documents, based solely on the knowledge and expertise of decision-makers and their technicians.

Rainwater recovery. Insufficient water in times of drought.

Pumping in the river and storage. The water is dirty when it rains. Outlying dwelling. The water is not used for drinking.

Donation of water by neighbours because the home cannot be connected to the communal network.

Water collected from neighbours. Communal project for the extension of a water network.

Fig. 4.8 Photographs taken as part of the "isolated areas" inventory, DASS-NC, 2022, photo credit: Authors

- That certain habitat areas located in "zones 3" are part of the drinking water supply projects of municipalities. Support for the implementation of all these projects would decrease the share of households experiencing water shortage from 0.9% to 0.4%, thereby allowing a reduction of the health risk associated with "lack of water" whilst also providing a response to the PEP objectives.

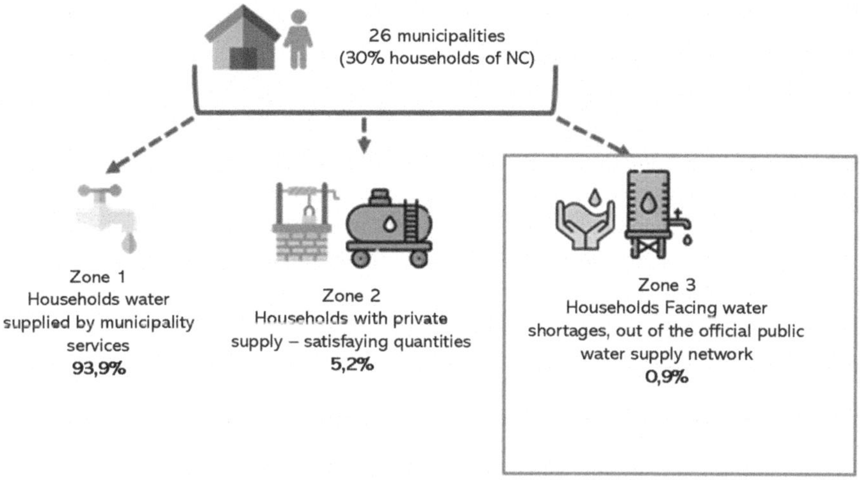

Fig. 4.9 Initial results from the "isolated areas" inventory in New Caledonia, DASS-NC, 2022

4.5 Future Policies and Perspectives for Drinking Water

4.5.1 Update of Regulation

In order to update the legal framework for drinking water and consider overall health risks, several projects will be implemented by New Caledonia in the coming years, including: (i) the definition of quality thresholds for the resource and drinking water standards for distribution, (ii) the creation of sanitary authorisations for resources intended for drinking water supply, (iii) the implementation of sanitary controls and the communication of their results.

Once these regulations have been established, municipalities and other providers will be responsible for bringing their facilities into compliance and rethinking the management of their networks. Any operator of a water system will then be required to monitor the distributed water, which was not previously the case.

This represents an upheaval in modes of operation, and the budgetary impact associated with these regulatory changes will nevertheless have to be thoroughly evaluated in order to allow the new paradigm to become firmly established and lasting.

Although essential, updating the regulatory framework for drinking water will not be the only work to be carried out to control the overall risk and preserve the health of Caledonians. It will absolutely have to be supplemented by monitoring mechanisms and in particular by the establishment of health indicators.

4.5.2 The Implementation of Water-Health Indicators, Serving as Surveillance Sentinels

Human health is the result of a complex balance between several genetic, social, economic and environmental factors. Improved knowledge and recent environmental health crises have raised awareness of the importance of interactions between the quality of the environment and the health status of populations.

These considerations have also stimulated, in the context of the PEP, the need to have relevant water-health indicators to strengthen health surveillance and allow public authorities to act in a way adapted to the realities of the populations and the Caledonian territory.

Acting as health surveillance sentinels, these water-health indicators will complement future regulatory provisions and will allow:

- development of priorities and action strategies in the field of water and health,
- monitoring of the effectiveness of action plans and programs,
- the general public to be informed by considering it as the main player in their health.

For the time being, the first indicators identified are derived from data collected by the Department of Health and Social Affairs of New Caledonia (DASS-NC) within its health jurisdiction. Diseases subject to mandatory reporting, non-specific surveillance, the sentinel network of physicians, biological data from medical analysis laboratories – these are all data that will be utilised to derive relevant indicators such as the percentage of reported cases of acute gastroenteritis, the incidence of waterborne diseases such as legionellosis or amoebiasis, the incidence of certain cancers, and the number of sales of medications intended to treat disorders associated with water quality, etc.

Indicators derived from data generated outside the DASS-NC will necessitate the establishment of a dedicated collection system and the appointment of a network moderator to ensure ongoing data collection with partners.

The water/health indicators project was initiated by the DASS-NC in 2019 and will be further developed in the coming years.

4.5.3 Towards a Change in Water-Related Behaviours

The behaviour of users now represents one of the most significant threats affecting New Caledonia's water resources.

The diagnosis established by the inter-departments water mission during the PEP highlighted drinking water uses and consumption patterns that are incompatible with the sustainable management of a water service, such as:

- overconsumption and waste (up to 690 L/person/day recorded in some municipalities compared to an average of 160 L/day/person in metropolitan France),
- non-payment for water services.

In the face of these alarming findings, concrete action was urgently needed. The very first water/health barometer established in 2020 by the DASS-NC and financed by the 11th EDF, responded to this need to assess the problems, attitudes and behaviours of Caledonians toward water, in order to identify key opportunities, communication strategies, and awareness-raising initiatives, as well as developing appropriate tools to encourage New Caledonians to adopt a more responsible approach to tap water management.

The results of this first barometer highlighted the representations, perceptions and attitudes of Caledonians in relation to the resource and water services, but also in relation to the health risk. At first glance, the results obtained often conflict with the data of water managers and the realities on the ground. This reflects the need to communicate, raise awareness and take more action on water issues in order to reduce existing gaps; this is especially the case given that Caledonians declare themselves in favour of changing their drinking habits. As a matter of fact, 65% are aware of the issues and are ready to change their habits. 35% are not informed or skeptical about the necessity to change.

Behavioural change aimed at improving an individual's health involves a voluntary process of relinquishing unhealthy habits in favour of adopting healthier behaviours for both individual and collective well-being. To achieve this goal, it is necessary to know which processes to choose and above all, how to maintain the "healthier" behaviour long term once adopted.

Behavioural change programs and associated communication strategies therefore encompass a wide range of activities and approaches, focusing on the individual, the community and environmental influences on behaviour and practices (cf. Chap. 9).

With this vision in mind and based on the barometer results, the DASS-NC aims to facilitate workshops to collaboratively develop tailored tools with various stakeholders in local communities to address field-specific issues and consumer profiles.

4.5.4 Anticipate Future Challenges

By 2025, according to the World Health Organization, half of the world's population will live in water-stressed regions. Therefore, in order to anticipate this situation, alternative supply methods and land use planning must be areas of consideration moving forward. Without a radical paradigm shift in current water management, health and water security in some regions will be undermined.

In health, anticipating future challenges also requires the development of One Health (Zinsstag et al., 2021). This concept emerged in the early 2000s, driven by the recognition of the interconnectedness between human health, animal health, and the overall ecological state of the planet. It aims to promote a multidisciplinary and comprehensive approach to health issues that involves close collaboration between the human, animal and environmental health sectors, as well as other stakeholders such as researchers, policy makers and civil society. This intersectoral approach is essential in a context where the origin of many diseases goes beyond traditional boundaries between disciplines.

In New Caledonia, environmental health has been integrated into public health since 2001 with the creation of the Environmental Health Unit and the awareness that the environment where you are born and raised could have an impact on your health. Environmental health promotes the preservation of the quality of life and the health of individuals by ensuring the quality of their surroundings. To this end, water is recognised as a key factor in population health and is addressed by the department in conjunction with human health. It is **in** this perspective and in the continuity of One Health, that DASS-NC has been involved in the development of the PEP and works daily to achieve its objectives.

Despite the growing recognition of the relevance of a One Health approach, its implementation still faces several challenges, including differences in priorities between the various health sectors, regulatory and institutional obstacles, difficulties of coordination between the various stakeholders, and so on. However, and even if in New Caledonia One Health is still in its early stages, the desire to better integrate the different aspects of health protection is progressively becoming ingrained in people's mindset and actions. After the COVID-19 crisis experienced by the country, cross-cutting work seems indeed to have become an obvious choice in the response to health problems encountered in Caledonian territory.

References

Cordier, S., Barguil, Y., Dominique, Y., St-Jean, A., Mengant, S., Le Bot, B., & Ayotte, P. (2017). *Niveaux d'imprégnation et déterminants de l'exposition humaine aux métaux en Nouvelle-Calédonie. "METEXPO" program*. CNRT "Nickel & Its Environment". 240 pages.

Faberon, F. (2017). Do Kamo, plan de santé calédonien. In 101 mots pour comprendre la santé en Nouvelle-Calédonie, A. Leca, F. Vangheluwe et F. Faberon (dir.), *Centre de documentation pédagogique de Nouvelle-Calédonie, Nouméa* (pp. 84–85).

Thery, G., Juillot, F., Meyer, M., Quiniou, T., David, M., Jourand, P., Ducousso, M., & Fritsch, E. (2023). Wildfires on CR-rich Ferralsols can cause freshwater CR(VI) pollution: A pilot study in New Caledonia. *Applied Geochemistry, 148*, 105513, 10 p. https://doi.org/10.1016/j.apgeochem.2022.105513

Zinsstag, J., Schelling, E., Crump, L., Whittaker, M., Tanner, M., & Stephen, C. (Eds.). (2021). *One health: The theory and practice of integrated health approaches*. CABI.

Chapter 5
Water Law and Legal Pluralism in New Caledonia

Victor David

Abstract The legal framework of New Caledonia on water management is still highly incomplete and reveals the imperfections of an unsatisfying conception of legal pluralism. After describing the current legal framework of New Caledonia, this chapter, through an historical analysis of the place of Kanak people regarding water law and the resurgence of endogenous law (or customary law) analyses the new legal water framework proposed by the government.

Keywords Water law · New Caledonia · Kanak · Legal pluralism

5.1 Introduction

The legal framework of New Caledonia on water management is still highly incomplete and reveals the imperfections of an unsatisfying conception of legal pluralism. As in other territories across the world in similar historical circumstances, during the colonial period, ancestral endogenous water-related regulations of the indigenous populations were set aside to implement imported water law from France. Since the 1990s, as New Caledonia entered an era of a step-by-step decolonisation process and the local Kanak people and their culture regained some authority, and together with land reforms, things began to change. Under the Matignon (1988) and Nouméa (1998) Agreements, the introduction of consociative institutional arrangements paved the way for a form of legal pluralism. Today, whereas public authorities apply public law to freshwater on public lands and water on private estates is

V. David (✉)
IRD (French National Research Institute for Development), Institut Méditerranéen de Biodiversité et d'Écologie (IMBE Marseille), Marseille, France
e-mail: Victor.david@ird.fr

C. Lejars, S. Bouard (eds.), *Water Policy in New Caledonia*, Global Issues in Water Policy 32, https://doi.org/10.1007/978-3-031-88196-1_5

mostly subject to property law[1] and easements, water present on indigenous customary lands was excluded from the public river domain according to Article 44 of the Organic Law of 1999[2] (Massenavette, 2011). Customary authorities manage the water resources present on their lands (springs, watercourses, and groundwater) according to their own customary practices. While we can only welcome the restored political dignity of the Kanak people and the resurgence of their customary rules, we cannot help but be disappointed by this compartmentalised approach to water law, which, even more so in the context of climate deregulation, transcends divisions of any kind. In the current context of endowing New Caledonia with a renovated legal framework through a participative process, it seemed timely to us to suggest a more inclusive form of legal pluralism, where the indigenous Kanak viewpoint and participation in decision making in a context of rebalancing is treated as equal to any other viewpoint and governance mechanism. Only then could we develop a different approach to water, seeing it not only as a resource or commodity as inherited from Roman (and French) law but consider it also, through Kanak eyes (by implying them directly in the making of law), as sacred, respected, and non-appropriable as it is the case in the legal systems of other plural societies, that sometimes even grant water the status of legal personhood. After describing the current legal framework of New Caledonia, this chapter, through an historical analysis of the place of Kanak people regarding water law and the resurgence of endogenous law (or customary law) considers the new legal water framework proposed by the government as a missed opportunity to achieve a much needed, true legal pluralism.

5.1.1 The Recent Recognition of Kanak Interests in Water Law in New Caledonia

Interesting developments have occurred since the late 1980s in the general institutional framework in New Caledonia allowing, after 150 years of colonisation, for indigenous Kanak customary rules and practices to have a say in the management of natural resources, including water.

Induced by the progressive decolonisation process initiated by the Matignon Agreements, the foundation of this evolution is to be found in a mix of reparative and distributive justice and an effective political participation of Kanak leaders in law making. However, there exists ample room for improvement: the current system still works in a silo form of legal pluralism that we need to move away from in order to create a more inclusive water governance.

[1] Both public and property laws in New Caledonia are either a mere copy of or largely inspired by the corresponding laws in mainland France.

[2] Loi organique modifiée N°99-1999 du 19 mars 1999 hereafter mentioned as 1999 Organic Law available at: www.legifrance.gouv.fr/loda/id/JORFTEXT000000393606

The origin, of course, lies also in earlier land reform policies started in the late 1970s to answer growing demands from the Kanak people and continued during the following decades with several ups and downs. From 1978, the "Plan Dijoud" (taking its name from the Minister for overseas territories in Paris) aimed "to resolve the land issue with a view to respecting the historical rights of the Melanesian communities".[3] In the "*Long-term economic and social development plan for New Caledonia*", presented to the Territorial Assembly in December 1978 in Nouméa by Paul Dijoud, land reform was indeed a key element. It advocated for returning traditional land to clans that had been confiscated from their ancestors since the mid-19th century and that were essential for the preservation of customary practices. The Territorial Assembly adopted a regulation (Deliberation No. 116 of 14 May 1980)[4] that defines the primary objective of the reallocation of land as addressing the "specific needs of Melanesian communities". The Territory did buy lands, but the reallocation to the Kanak was minimally effective. A second wave of land reform was initiated by the socialist government through an ordinance law of October 1982. The Prime Minister's report to the French President presenting the ordinance is worth a read as it correctly analyses the colonial heritage but does not propose audacious solutions to erase past errors:

> ***The land problem in New Caledonia stems from the unequal distribution of land between Melanesian populations and those of colonial origin.*** *[…]*
>
> *The Melanesian community's particular conception of the land, around which all social relationships are organised, the notion of individual appropriation being excluded, makes it even more unacceptable, in its eyes, that European private property represents 400,000 hectares, while the reserves, within which Melanesian methods of tenure are still in use, amount to just 160,000 hectares.*
>
> ***Generally speaking, the problem of land tenure is at the root of the poor integration of the Melanesians into the multi-ethnic society in New Caledonia.***
>
> *It is therefore necessary to prevent this situation, where Melanesian claims culminate in land occupations, from leading to a confrontation between communities with profoundly different conceptions of the law, and to remove the obstacle that the land issue poses to their relations.*
>
> ***To this end, the proposed arrangement recognises that vis-a-vis European-type established rights, customary rights of a different nature may pre-exist. This recognition is a duty to the Melanesian community.***
>
> ***The principle of the ordinance is therefore to recognise the special rights of Melanesians,*** *while safeguarding ownership.*[5]

Contemporary Kanak water rights on their customary lands can therefore be traced to these land reforms from forty years ago even though explicit jurisdiction on water was only established in 1999 as detailed below.

---

[3] See ADRAF. https://www.adraf.nc/la-reforme-fonciere/historique#entre-1978-et-1986.

[4] https://www.adraf.nc/images/sampledata/terres-coutumieres/deliberation-116-du-14-mai-1980.pdf.

[5] Ordonnance n° 82-880 du 15 octobre 1982 relative à l'aménagement foncier, à l'établissement rural et à la reconnaissance des droits coutumiers sur le sol de la Nouvelle-Calédonie et dépendances. https://www.legifrance.gouv.fr/loda/id/JORFTEXT000000339928/. Our translation.

It is a generally a shared view that while water is an abundant resource in New Caledonia, it is unevenly distributed in space and time. It is also a vulnerable resource, as it is threatened by a number of hazards, including mining and agricultural pollution, invasive species, and over-use.

Since the 1990s and until recently, water management in New Caledonia and a clear and precise division of powers between local authorities in this area lacked any real guidelines and only had a poorly developed legal framework. As a matter of fact, we had an outdated legal framework: one local regulation (Deliberation No. 105 of 9 August 1968 of the Territorial Assembly and other property law provisions of the French Civil Code that are now part of the New Caledonian Civil Code. Other regulations taken for the needs of public health also exist.

Since the Matignon-Oudinot Agreements in 1988, and even more so with the Nouméa Accord of 1998 and its legislative translation, the amended Organic Law No. 99-209, an internal federalist-inspired mechanism was introduced at two levels. Firstly, between the State and New Caledonia and its three provinces (Loyalty Islands, North and South), and secondly between New Caledonia and its three provinces. New Caledonia and its three provinces have been granted deliberative and executive institutions, regulatory autonomy, and shared areas of jurisdiction. As mentioned in the introduction, since 1999 this framework divides authority for managing water and aquatic environments into four different levels of local governance:

(A) New Caledonia, which manages water located outside customary lands; (B) customary authorities, which manage water resources located on customary lands; (C) provinces, whose environmental jurisdiction includes aquatic environments, and the North and South provinces have requested and obtained, as permitted by the 1999 Organic Law, the delegation of responsibility for the management of water use authorisations and the maintenance of watercourses; (D) municipalities (or their groupings), which are responsible for supplying the population with drinking water, as well as for wastewater treatment.

Let us for instance focus on the division of water resources on a place-based approach, with customary owners managing water on their lands, i.e. land belonging to the Kanak people in accordance with the Nouméa Accord. According to Section 44 of the amended Organic Law No. 99-209 of 19 March 1999 on New Caledonia, the resources located on these lands, including rivers, lakes, groundwater, and springs, are not part of the public waterway domain of the State, New Caledonia, or its provinces. This therefore excludes them from New Caledonia's public river domain and associated legislation. Customary lands represent 27% of the surface area of New Caledonia and house 52% of drinking water catchments. On these lands, chiefdoms and clans exercise sovereignty over their territory and water, like land and natural resources, is an element of common heritage and cultural identity. Any project to install water networks is subject to the prior agreement of the chiefdom and the clans through the mechanism of the "*palabres coutumiers*" that is usually granted on the basis of the results of an assessment of the environmental and socio-cultural impacts. As a result, one might think that these water resources are governed by the customary rules in place. First, we need to say that the rationale of this indigenous jurisdiction on water on customary lands is not very clear and was probably not properly considered in 1998. To use Elisabeth Macpherson's conceptual tools (Macpherson, 2019), we do not think it was intended as giving water rights and jurisdiction - in par with formal authorities - over water, in line with the Kanak cultural relationship with water. That the national legislator in 1999, respected the Preamble of the Nouméa Accord that called for the reparation of decades of land seizures in the colonial period is evident. That the same legislator did want the Kanak to have their way as *owners*—in the sense of French property law - on their recently recovered lands is also evident. We do not believe it was intentionally meant to introduce legal pluralism in water management[6] as the planned new framework shows.

5.1.2 Limits of the Proposed New Legal Framework

The global and local issues surrounding water as a precious resource, with a human right of access to water, and the need for New Caledonia to adopt a policy in this area had already been highlighted in a resolution adopted by the Economic and Social Council (CES) of New Caledonia in 2006: the CES referred to an insufficient and obsolete regulatory system and a dispersal of resources and actions. These observations were reiterated at the Water Conferences organised at the initiative of the CES in 2008 and 2009 but were not really considered in terms of public policy. The issue of water policy returned to the forefront in 2018 with the organisation of the first Water Forum (Forum H_2O) in New Caledonia, initiated by the government, bringing together, as in 2008, a large number of players from civil society, international experts, local researchers, and the political and customary authorities of New Caledonia. Various working groups and consultation meetings at local level, in the three provinces, took place in the second half of 2018 and led in March 2019 to the adoption by the Congress of New Caledonia[7] of a "*Schéma d'orientation pour une Politique de l'Eau Partagée*" (Schéma PEP hereafter) for the country.

The Schéma PEP reviews the many social, economic, and environmental factors and problems that have combined to demonstrate the crucial need to establish a country-wide water policy. The making of the Schéma PEP did associate customary authorities but it clearly stops close of real innovation in terms of legal pluralism in water management. The Schéma PEP adopted in 2019 by the Congress of New

[6] Actually, little attention was paid at first to this self-management of their water resources recognised to customary authorities until difficulties related to catchment areas for general purpose arose.

[7] Congrès de la Nouvelle-Calédonie. Délibération n° 419 du 19 mars 2019 portant approbation du schéma d'orientation pour une politique de l'eau partagée de la Nouvelle-Calédonie. https://davar.gouv.nc/sites/default/files/atoms/files/deliberation_ndeg_419_du_19_mars_2019_portant_approbation_du_schema_dorientation_pour_une_politique_de_leau_partagee_de_la_nouvellecaledonie.pdf

Caledonia[8] is based on a very technocratic approach to water issues and there is no room to acknowledge local populations' perceptions or relationships with water. The Plan's six strategic priorities indeed neglect legal and cultural pluralism : "*(1) protect catchment areas and strategic resources, preserve the environment; (2) provide 150 L of drinking water per day per inhabitant; (3) raise awareness, communicate, inform, educate and train: make all New Caledonians "hydro-eco citizens"; (4) move towards "zero discharge of untreated water"; (5) put water at the heart of all planning, housing and economic development projects; (6) better manage water to increase local agricultural production and achieve a food coverage rate of 50%.*"

The three cross-cutting priorities of New Caledonia's shared water policy, as defined by the Schéma PEP, are as follows: "*A. Establish a legal framework and a police force suited to an effective water policy; B. Improve knowledge to better protect, preserve, plan and manage; C. Develop the skills of all those involved in the water sector to improve performance.*"

The New Caledonia government presented a bill in 2023[9] (the Water Bill hereafter) to implement the first strategic priority we just mentioned. Built on the participation based Schéma PEP and after months of consultations the proposed bill can be considered as a progress as it recognises Kanak water rights and introduces a number of procedures to involve customary authorities, for instance their approval is needed to access catchment areas situated on their lands. The bill implicitly acknowledges, in compliance with the 1999 Organic Law, that it will not apply to water present within customary lands, thereby paradoxically introducing a system of dual standards. On the one hand it states that water is the common good of all Caledonians and that everyone has a right to water usages. On the other hand, this statement does not apply to water in Kanak customary lands. This oddity was recalled by the CESE[10] of New Caledonia in its opinion[11] on the water bill: "*The [CESE] was particularly struck by the fact that only public land is concerned by this text, as it is clear that rivers do not stop at the boundaries defined by public law or custom. Pollution or misuse upstream or downstream has an impact on the whole river. It therefore fails to understand how we can claim to watch over water resources and define them as a common good when the text de facto excludes a large part of New Caledonia's land*".

This observation by the CESE seems all the more relevant when we look at Section 54 of the Water Bill which reads as follows: "*A deliberation prohibits,*

[8] See note 8 above for all quotations from the Schéma PEP. Our translation.

[9] See: "Projet de loi du pays relative au domaine public de l'eau de la Nouvelle-Calédonie et à la protection de la ressource en eau (rapport n° 37/GNC du 1.05.2023 » at https://www.congres.nc/textes/projet-de-loi-du-pays-relative-au-domaine-public-de-leau-de-la-nouvelle-caledonie-et-a--la-protection-de-la-ressource-en-eau-rapport-n-37-gnc-du-31-05-2023/. Our translation).

[10] The Conseil Economique et Social of New Caledonia we mentioned earlier became the CESE in 2013 by the adding the Environment to its jurisdiction.

[11] CESE. AVIS n° 32/2022 du 1er décembre 2022 concernant le projet de projet de loi du pays relative au domaine public de l'eau de la Nouvelle-Calédonie et à la protection de la ressource en eau. Available at:https://cese.nc/sites/default/files/avis-et-voeux/2022/AVIS%2032-2022%20CESE%20NC.pdf.

regulates or requires authorisation from the Government of New Caledonia for spreading, dispersal and infiltration activities likely to alter ***the quality of water belonging to the public water domain***." We share the doubts raised by the CESE about water within customary lands.

Sections 52 and 53 explore the necessary cooperation between the New Caledonia government and customary authorities for the shared management of catchment areas and Part III of the Water Bill (Sections 55 and 56) deal with the content of agreements that should be put in place for the shared management of watercourses, lakes, springs, and groundwater on customary land.

As we can see, the Water Bill proposes nothing but two different regimes, one it defines for water in public lands, leaving to customary authorities the freedom to do as they please on their lands with the extra guarantee that the government cannot expropriate customary landowners in the general interest. Who can complain about it? We have come a long way. In 2001, when the CES published a progress report on freshwater in New Caledonia, the only Kanak they interviewed were two administrative officers of the Loyalty Islands. In the 2008 Water Conference (Assises de l'eau) no customary authority (not even a Customary Senator) was invited to deliver a speech about water. In the Conference summary published by the CES in 2008, there is not even an allusion to Section 44 of the 1999 Organic Law on water on customary lands. Reference is made to the 1968 regulation which gives full jurisdiction to New Caledonian authorities ("watercourses, lakes, springs and groundwater belong to New Caledonia").[12]

Yet, though the process to build the new Shared Water Policy had clearly integrated customary authorities and stakeholders, the recent Water Bill was another missed opportunity to really build a common destiny around water management in New Caledonia, in particular in the context of Climate Change.

5.1.3 A Missed Opportunity to Achieve Legal and Cultural Pluralism

Climate Change is now a crucial, if not the most significant, factor to consider when drafting legislation on water governance and requires a renewed and shared vision to protect what is already the most precious resource on Earth. Water will be the foundation of adaptation policies in the near future. Whether the proposed water law in New Caledonia will stick to this vision is yet to be seen. For the moment, with the coexistence of formal and customary rules in a plural territory, it is reminiscent of ancient plural Roman society: populations of different origins, citizenships, social classes, and territories, united under the banner of Rome, with different access to water and different needs, and a variety of rules and practices around the use and management of water resources. But it should not be forgotten how in a few

[12] CES. Synthèse des Assises de l'Eau. 2008. p. 15. https://cese.nc/sites/default/files/publications/files/2017-07/16088714.PDF

centuries the Roman Empire put aside these parallel rights and practices by substituting them with the *ius civile* it was promoting. The situation is hardly any different in today's plural post-colonial societies, like New Caledonia, where ancestral rights and modern rights continue to battle it out, with the latter more often than not attempting to ignore or replace the former. Pluralism should be both the source of and the guarantee of a shared vision on water governance, or in other words, an inclusive one. On the other hand, we feel it necessary to qualify the idea of an exclusive (akin to apartheid) pluralism: both "modern" water law and customary law must be neither demonised nor idealised. Accepting the need for legal pluralism in water matters means taking several facets of the problem into account and moving away from conceiving water law either as ancestral rules that cannot be challenged no matter what, or only as a set of binding public policies and legal standards established by engineers and technicians, backed up by penal sanctions and civil redress mechanisms, adopted by institutions with a legitimate monopoly in a state with rule of law on the production and implementation of law.

For instance, we need to:

(a) Recognise the existence of non-State actors in the field, and practices and rules, often local, concerning the use of water, which are often ancestral and have survived the passage of time and attempts at assimilation.[13]
(b) Recognise the existence of (1) rights *to* water held by various stakeholders to satisfy a number of vital, everyday needs (human and animal), or those required for economic (mining) or agricultural (irrigation) activity, or even those corresponding to a certain concept of life and leisure (watering golf courses, for example); (2) Water rights based on ownership (private, public, communal) or non-ownership (sacred water, water as a legal entity); (3) The economic rights of indigenous or local populations to water resources.

In other words, it is important for a water law, in a context of cultural, inclusive pluralism to consider the elements mentioned in (a) and (b) above, and to translate them into objective rules. It turns out that this is rarely the case. Most "modern" water law—and it is the case of the NC water bill—continues to be drawn up, based on theoretical models, liberal ideologies, and the promotion of technical progress. It aims to forge uniformity or modify behaviours, and in post-colonial states, especially in the presence of a multicultural society, can be far removed from the realities on the ground, and a source of perpetuated inequalities and injustices. This post-colonial discrepancy is a source of confusion: on the one hand, if not outright negation or rejection, endogenous law and practices are discredited as obsolete and incompatible with the needs of the modern world; on the other, we have a formal law whose social acceptability nobody has bothered to check, and which is therefore often largely ineffective. In the name of the "general interest", which reinforces

[13] These rules are usually adapted to territories that often have their own system of resource allocation based on needs, the interest of all and solidarity rather than ownership, and their own system of conflict resolution. These rules and practices may constitute what is generally referred to as customary or endogenous law.

development policies, formal institutions can make decisions on water resource management that ignore or erase local and/or ancestral rules. The ignorance by post-colonial authorities of Kanak water relations up until the last decade is a compelling testament to this fact.

The 2023 Water Bill in New Caledonia envisages a situation where formal and customary law coexist and compete to regulate access to and use of water on property law basis. We know that it is competition that generates frustration and misunderstandings.

However, despite certain prejudices, water law—like law in general—is by no means monolithic and is never set in stone... except for a time! It can vary over time according to the political options of those in power, and the technical nature of solutions for accessing, transporting, and using water, with engineers playing a dominant role. It can adapt to changing times and climate, needs, pressure from lobbies, budget deficits, environmental changes, and so on. Contemporary formal law has a duty to consider various social, technological and, above all, environmental changes that endogenous law may not be able to manage, as it has never been confronted with these situations. Several remedies to an overly top-down approach, initiated by international law (Principle No. 10 of the 1992 Rio Declaration, for example), now also allow law to be co-constructed, on a more participatory basis in other countries.

More recently, water law has had to adapt to the recognition, by other branches of national or international law, of the rights of indigenous peoples, and indeed of endogenous or customary law. For more than half a century, international law has been progressively advancing the position of indigenous peoples, their rights and their traditional knowledge in "modern" societies and has made a major contribution to changing the situation in this area. These include, of course, Convention 169 of the International Labour Organisation (ILO) (1989) and the Rio Convention on Biodiversity (1992). More recently, the Paris Climate Agreement (2015) underlined the essential role of traditional knowledge, not forgetting the 2007 United Nations Declaration on the Rights of Indigenous Peoples. Over the last few decades, many national legal systems have enshrined the reaffirmation of endogenous law within the communities that make use of it, paving the way for situations of legal pluralism. In the wake of these developments, numerous demands for recognition of ancestral rights to water have been made on every continent, with sometimes mixed results, and other times spectacular ones, in that they have transformed the legal status of water, as we shall see later.

Regarding endogenous law, some precautions are needed too.

The enshrinement of endogenous law in contemporary legal systems, out of respect for the peoples who developed it and continue to apply it, is often the basis of contemporary legal pluralism. The same respect for customary law and traditional knowledge has led to the development of various forms of 'community management' of natural resources, coastal areas, and water, of course, all over the world. However, it is important to avoid idealising customary law and, above all, isolating it from society, either as the only legitimate system or as a substitute for formal law that is presupposed to be or is genuinely inadequate or ineffective. Professor Spiertz reminds us of this in his work on legal pluralism in the field of water, in his study of a rural society in Indonesia:

> *Colonial legal anthropology and part of post-colonial legal anthropology have continually conceptualised (customary) law and practices independently of their socio-legal, administrative and political environment more generally. Conceptual rigidity compounds this by preventing the processes of change within these systems from being perceived and taken into account* (Spiertz, 2000).

These misconceptions have often led to the mystification and romanticisation of "*the ancient and deep-rooted rules and practices of indigenous communities, based on stable, equitable and conservation-oriented local knowledge*" (Spietz, 2000).

Not all indigenous and traditional communities are based on equity and may harbour multiple ancestral inequalities and discriminations that can lead to differentiated access and rights to water. Not all traditional practices have the direct aim of conserving and protecting nature, but our objective here is not to venture into the realms of essentialism or cultural exceptions.

Another limitation is that customary law is neither a uniform or structured system but is diverse in many respects. As Spiertz reminds us, observers from outside traditional societies or nationals of traditional communities educated in the universities of colonial and post-colonial metropolises have been able to evoke customary law in the image of formal and state law, as a coherent legal order that can replace formal law in its permanence and its vocation to apply to an entire territory and to all communities. In reality, this is rarely the case. The example of New Caledonia, where individual customary chiefs claim 'sovereignty' within the limits of their 'customary zone of influence', is a good illustration of this in our view, as is the desire expressed by the customary senate to draw up a 'Common Base of Kanak Values' to meet an external demand for uniformity in customary law, in this case Kanak law.

5.2 Conclusion

We have seen how water law can reveal the existence of legal and cultural pluralisms, and how difficult it can be to reconcile different worlds.

For a simplified approach to the question, we can say that there are two situations:

(a) Legal orders operate in silos, with the delimitation between formal and customary law being territorial and/or based on (the claim to) belonging to a particular human group (a state or ethnic "citizenship" of sorts); each order ignores the other, and everything goes "smoothly" as long as there are no conflicts. In this scenario, the hierarchy between legal orders typically leans towards dominance of the formal legal order. This situation is one of suffered and rarely chosen apartheid. The 2023 New Caledonia Water Bill falls into this category.
(b) It could be different if a willingness to reconcile cultural diversities exists: the State recognises customary law as part of its formal legal order, and collabora-

tive mechanisms and institutions are set up to reciprocally recognise the legitimacy of the legal orders involved.

To be successful, this solution implies that formal law does not impose itself, and that consideration for customary law and traditional practices is not merely cosmetic. It also assumes a level of decentralisation as necessary, particularly where the size or layout of the country demands it.

In the case of renewing the water legal framework in New Caledonia a different solution than the proposed Water Bill was (is?) not impossible both in terms of legal and cultural basis.

The exclusion of water within customary lands from the legal regime of public waters was nothing but a mechanical application of Section 18 of the 1999 New Caledonia Act which says that “Customary land and property located thereon belonging to persons with customary civil status are governed by custom.” It was combined with Sect. 44 of the same legislation that mentions that New Caledonia’s public domain “also includes, subject to the rights of third parties and except where they are situated on customary land, watercourses, lakes, groundwater and springs.”

It is therefore essential to reach an agreement between the elected and administrative authorities and the customary authorities in order to establish optimal water management, whether we remain within the concept of water as “property” or as “common good” or whether we agree to consider the symbolic views of the Kanak people. In recent years, in countries where endogenous and exogenous populations and/or legal orders coexist, initiatives have been undertaken to try to bring together systems that have remained alien to each other. Voices, particularly those of indigenous peoples, in New Zealand and in the Americas for instance, have been heard, both to assert their legitimacy to be associated with the management and development of water resources and also request that water be considered differently, with more respect. These voices invite us to recognise the interdependence of the elements of nature and have led governments and judges from several parts of the world to recognise among other natural elements, rivers as living legal entities. In June 2023 the province of Loyalty Islands in New Caledonia introduced in its protected species regulation[14] the legal category of Natural Legal entities and included sharks, and marine turtles as the first to be listed in this category. Departing from the traditional classification inspired by Roman law between objects and subjects of law, this new category is the proof that it is possible to draw on the wealth of legal and cultural pluralisms and crossbreed the law: Elements of Nature are neither objects nor legal persons but legal entities, having their own rights and standing, with proper legal regimes. Considering watercourses and water sources as legal entities would open new avenues of cooperation between customary authorities and the New Caledonia government and its services.

[14] Délibération n° 2023-28/API du 29 juin 2023 relative au code de l’environnement de la province des îles Loyauté. *Journal Officiel de la Nouvelle-Calédonie* N° 10602, dated July 18, 2023. p. 14804. This deliberation was however challenged in court by the State on the ground of incorrect provincial jurisdiction and annuled in July 2024.

References

Macpherson, E. J. (2019). *Indigenous water rights in law and regulations*. Cambridge University Press. (See Part 1 in particular).

Massenavette, C. (2011). La reconnaissance législative du domaine fluvial de la Nouvelle-Calédonie : la fin d'une longue période d'insécurité juridique. *Revue juridique de l'environnement, 36*(1), 17–30.

Spiertz, H. J. (2000). Water rights and legal pluralism: Some basics of a legal anthropological approach. In *Negotiating water rights* (pp. 162–199). London: Intermediate Technology Publications. p. 9.

Part II
From Water Representations to Water Policy: A Look Insight Customary Lands

Chapter 6
Water Tensions and Social Transformations on Customary Lands

Caroline Lejars, Séverine Bouard, Charline Nekiriai, and Catherine Sabinot

Abstract In New Caledonia, in the Kanak communities, "water is the link", the link between Heaven and Earth, between the Mountain and the Sea, but also between men and between clans. This article aims to show how the evolution of water networks management, from the "traditional" irrigation canals to the drinking water supply network, has transformed and still transforms uses of water in tribes and how these changes modify the links between men and tribe social organization. In the specific context of New Caledonian negotiated decolonization, focusing on techniques and technical changes around access and uses of the water show the impact of colonization and mining development on the management of the resource and on customary rules.

Keywords Hydraulic systems · Hydrosocial cycle · Irrigation canals · Decolonization

---

This chapter has been previously published in French in the review 'Développement Durable et Territoire'. The review agrees for a translation in English and publication in this book. Lejars C., Bouard S., Sabinot C., Nekiriai C., (2019). Quand "l'eau, c'est le lien": Suivre l'évolution des réseaux d'eau pour éclairer les pratiques et les transformations sociales dans les tribus kanak, Développement durable et territoires, Vol. 10, n°3.

C. Lejars (✉)
CIRAD (French Agricultural Research Centre for International Development), UMR GEAU (Univ Montpellier, CIRAD, INRAE, IRD, BRGM, AgroParisTech, Institut Agro), Montpellier, France
e-mail: caroline.lejars@cirad.fr

S. Bouard
IAC (New Caledonian Institute of Agronomy), Pouembout, New Caledonia
e-mail: severine.bouard@iac.nc

C. Nekiriai
IAC (New Caledonian Institute of Agronomy), Nouméa, New Caledonia

C. Sabinot
IRD (Research Institute for Development), UMR Espacedev (IRD, Univ Montpellier, Univ Guyane, Univ La Réunion, Univ Antilles, Univ Nouvelle Calédonie, Univ Perpignan), Centre IRD, BP 05, Nouméa, New Caledonia
e-mail: catherine.sabinot@ird.nc

C. Lejars, S. Bouard (eds.), *Water Policy in New Caledonia*, Global Issues in Water Policy 32, https://doi.org/10.1007/978-3-031-88196-1_6

6.1 Introduction

Despite the symbolic importance of freshwater in the Pacific and the attachment of island communities to their rivers and channels, few studies have explored the issue of freshwater from a historical, anthropological or sociological standpoint (Wagner & Jacka, 2018). Research into the relationships between societies and their environment has generally analysed the development of agrarian systems or relationships with the sea without exploring the question of rivers and springs or the taro plantation irrigation systems (Sabinot and Lescureux, 2019),[1] which nevertheless extended across vast areas on certain islands (Sand, 2016).

New Caledonia is no exception. In many Kanak communities,[2] "*water is the link*"; it is the link not only between Heaven and Earth, mountain and sea, but also between men and between the clans[3] (Nekiriai, 2017; Gouvernement de la Nouvelle-Calédonie, 2019). Echoing the statement of the great Kanak chief quoted at the beginning of the introduction, "*cutting the water supply means cutting the link*" (Nekiriai, 2017).

In light of these words, analysing the management and development of the water networks is the ideal means of shedding light on the social transformations at play in Kanak society. In this singular context, water and water networks are a "*mirror of society*", to quote Aubriot (2004), a reflection of social ties (Aubriot & Riaux, 2013). They are also a reflection of the links to the environment and to "nature".

Focusing on tribal management of water networks seems all the more relevant as the context in New Caledonia is one of negotiated decolonisation as stipulated by the Nouméa Agreement of 1998. The Nouméa Agreement enacted and organised the "decolonisation" of the archipelago, in particular by gradually transferring powers from the French state to New Caledonia, organising a self-determination referendum in 2018 and establishing "New Caledonian citizenship". Against this backdrop, the governance of water in New Caledonia is the result of a quite specific institutional organisation with an incomplete regulatory framework which is not entirely in line with local usages and practices. Customary lands have been excluded from the public waterways domain pursuant to Article 44 of the Organic Law of 1999, amended in 2009 (Massenavette, 2011). The customary authorities manage the water resources located on their lands (springs, rivers and groundwater). These lands account for 27% of the total area of New Caledonia and 52% of drinking

[1] Taro is a tropical plant in the Araceae family with edible tubers.

[2] The adjective and noun "Kanak" refer to the indigenous people of New Caledonia. In the 2014 census, 39% of all inhabitants of New Caledonia declared they were Kanak.

[3] Kanak society is built around its own customary structure consisting of (i) the family or lineage and (ii) the clan, which brings together all the lineages of the same stock, governed by a clan chief. This social organisation is overlapped by an administrative organisation resulting from the colonisation process consisting of (iii) the tribe or chiefdom, which incorporates several clans and is recognised as a traditional community by the decree of 24 December 1867, granting it legal status, and (iv) the district, an administrative creation more or less in line with the spheres of influence of each great chief.

water catchment (Davar, 2018). Within these lands, the chiefdoms and clans exercise sovereignty over their own territory. Like land and natural resources, water is an element of common heritage and cultural identity. Any project aimed at installing water networks is subject to the prior consent of the chiefdom[3] and the clans and is granted on the basis of the results of an environmental and socio-cultural impact analysis (Customary Senate, 2014).

While water is not a scarce resource in New Caledonia, tensions nevertheless exist, in particular on customary lands, with regard to "sacred" water (as certain springs or water holes are prohibited or taboo[4]; see Demmer, 2002, or Leblic, 2005, for example), the payment of water bills and the over-consumption of drinking water (Davar, 2018) or pollution, and in particular pollution resulting from mining activities (Lejars et al., 2016; Gosset et al., 2019). As within other indigenous populations of the Pacific, South America or Canada (Hidalgo et al., 2017; Seemann, 2016; Jackson & Altman, 2009; Hsiao, 2012), water and its associated infrastructures could become a source of grievances and conflict in New Caledonia (Trépied, 2004, 2011).

Based on a technical and historical analysis of tribal water networks, the article aims to shed light on the social transformations observed within the tribes and to provide a better understanding of the tensions relating to use of water. It focuses on tribal water networks from the taro plantations, which were traditionally irrigated by open channels, to underground water networks which have gradually replaced the irrigation channels. Echoing the works of Barbier and Trepos (Barbier & Trépos, 2007), this article analyses the role of these technical elements in the Kanak communities as well as the means by which they have contributed, and still contribute, to the changing social and political relations. It endeavours to shown how colonisation, followed by the development of the mining sector, have brought technical changes with regard to water and how these changes contribute to altering uses and practices linked to water, the spatial structure of the tribe, links to the resource and social relations within the communities and with the water managers. Focusing on technical elements also explains current water-related practices (in particular over-consumption) and the tensions caused by this resource.

The study focuses on two tribes which have been impacted by colonisation and the development of mining. The first, the Nekliaï tribe living in the municipality of Poya in the North-West, is the main subject of the study while the Mîa tribe, in the municipality of Canala in the South East, provides a secondary subject. Drawing on case studies, we will analyse (i) the changing water distribution and management methods in customary lands from the open irrigation channels to the implementation of underground drinking water networks first by the miners then by the municipalities; and (ii) the changes observed in the associated practices and uses. The starting point for the process and the basis for the interviews conducted on site was

[4]The notions of "sacred" and "taboo" are polysemous and dynamic (Leblic, 2005). The conditions of the restrictions and access to the sites differ from one individual or clan to another. Non-compliance with these restrictions results in sanctions which vary according to the sites and the people who violate the restrictions.

an analysis of how the irrigation channels work. In light of the strong symbolism associated with water and the channels, the process was also based on the idea of the hydro-social cycle of water (Budds & Linton, 2014), incorporating the symbolic, cultural and historic dimensions of water and the networks as well as the power plays involved in managing this resource.

In the first section, we develop the theoretical framework of the analysis and the survey methodology. After a historical reconstitution of how the networks operate and are managed, we then show how the technical and historical analysis reveals the social dynamics. Finally, we discuss the benefits of focusing on the technical elements to highlight the social transformations at work within the tribes, the power games at play between the tribe, the municipality and the mining companies and the tensions caused by the current use of water.

6.2 Method: Following the Networks to Decipher Social Transformations and Highlight Practices

6.2.1 Theoretical Framework: The Taro Plantations and the Underground Networks as an Entry Point to the Hydro-social Cycle

Through the notion of the hydro-social cycle of water, Budds and Linton (2014) attempted to move beyond the dualist categories of "water" and "society" by recognising water to be a joint production of nature and societies. Historically, the organisation of the Kanak tribes incorporates resource management as a defining element of social organisation (Sand, 2016). Kanak organisation focuses on lineages (families) endowed with a function, or in other words a "knowledge" or an "ability" resulting from genetic inheritance, revelation or gift. The clans are organised around these functions and united by solidarity which is essential to their survival (Bensa & Leblic, 2000). Some functions, such as water management which falls to the water clan, are so important that they impact on the entire clan or tribe. At present, even if many of these functions no longer exist, certain individuals remain who are useful to the community or have retained a mystical link with the "supernatural powers". Against this unique backdrop, drawing on the notion of a hydro-social cycle of water to understand social transformations seems particularly relevant.

In this article, we have used the open water channels traditionally used to irrigate the taro plantations and managed by the water clan as a point of entry to the hydro-social cycle. In many works, the material and technical dimension of water systems is at the heart of the analysis of the hydro-social cycle. These works present water and the associated techniques both as a power issue strongly linked to technical and traditional knowledge and as a key indicator of the social relations at work in a given society (Aubriot & Riaux, 2013).

On the one hand, they illustrate the importance of understanding the technical activity in order to grasp the link between water and society while also showing how the technical and social spheres interlink, as technology is a social act (see the works of Bédoucha, 1987; Lemonnier, 1980; Wateau, 2002). Descriptions of the spatial techniques and water management organisation provide grounds to examine the power games underpinning the technical choices (Obertreis et al., 2016; Janty, 2013; Récalt et al., 2013, Swyngedouw, 2004). With regard to the hydro-social cycle, technical devices clarify and make social and political dynamics visible (Fofack, 2018).

In light of the entrenchment between the social organisation and natural resource management in the Kanak tribes as well as the strong symbolism associated with water, we surmised that a technical and historical analysis of the taro irrigation networks would reveal the social transformations to be observed in Kanak society. In many tribes, these networks have gradually disappeared or been transformed into underground water networks. Traditionally responsible for managing the taro plantations, the water clan and the water master have also lost some of their functions and are now sometimes difficult to identify within the tribes. Visiting the field via the remains of the taro plantations and taking a detour via the "technical element" enabled us to meet and identify the clans and former water masters. Some plantations have indeed disappeared to a large extent and members of the water clan preserve the knowledge and know-how of how to construct water works and manage the channels that passed through the tribe. In addition to the knowledge and know-how of the water clan, the water masters from the water clan ensured that the rivers and works were respected. Despite the arrival of new water management actors (in particular the mines followed by the municipalities), the water clans, while no longer visible or involved in actual water management, retain the "memory of water", the historical knowledge of the resource and its management and, in certain cases, maintain a specific and symbolic link with springs and the river (often via their totem[5]).

Examining the disappearing taro plantations and the water clans seemingly offers a means of analysing the relationship that Kanak communities maintained, and still maintain, with water.

[5] A "totem" is an element of the natural world through which the ancestors of the clan manifest themselves. Each clan sees its ancestors appear in different forms (a thunder clap, a lizard, a shark, etc.) and, following the rituals performed to communicate with the dead, these manifestations of nature are seen as an answer from the ancestors. Ethnologists have called these manifestations of ancestors "totems" (Leblic, 2005), a term now used by Kanak society as a whole.

6.2.2 Study Site and Surveys Conducted

We focused our work on two areas (Fig. 6.1): a primary area involving the Nekliaï tribe in the north-western region of Poya, and a secondary area involving the Mîa tribe near Canala. In each area, irrigated taro plantations were managed collectively by the tribe, gradually falling into neglect. One of the two areas chosen was strongly affected by the implantation of nickel mines and the resulting pollution; the other was subject to more indirect impacts through the economic activity generated by the mines in the municipality, in particular with the development of wage-earning activities within the tribe.

The Nékliaï tribe is part of the municipality of Poya, living on land 4 km north-east of the centre of the village. The land is crossed by the Nékliaï River and is also home to the Nê Erhèji creek, which is the source of one of the catchments for the Poya municipality's drinking water network. According to the last census, conducted in 1996, the tribe totalled 194 residents, most of whom speak *ajië* and French. The Nékliaï formerly owned taro plantations irrigated by means of open water conduits. With the expansion of mining activities from the 1980s onwards, the taro plantations gradually disappeared. From the very beginning of nickel mining operations in the 1950s, the Nékliaï River was polluted by the activity of the Nétéa

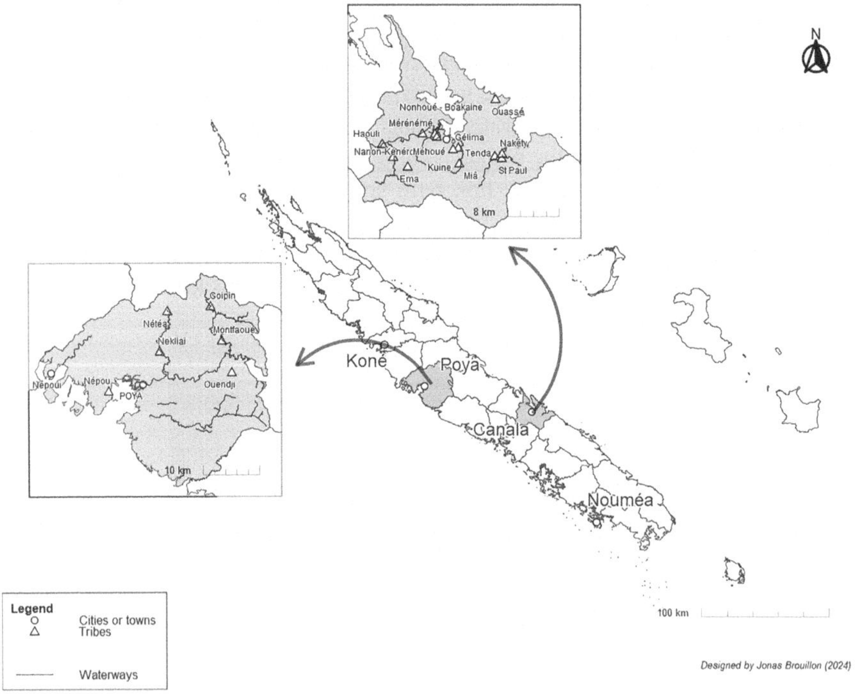

Fig. 6.1 Location of the study sites

mine, located upstream, as the waste and unusable residue from the nickel was dumped into the river. Although the mine closed during the 1970s, the population continues to complain that the river silts up[6] after every rainy spell.

The second area is that of the Mîa tribe, located 6 km from the village of Canala and at an altitude of 300 m, close to the River Xwê Ciü. According to the most recent census, the tribe totalled 125 residents who speak both French and *xârâcùù*. This tribe is known for its irrigated taro plantations, some of which are still watered every year during the water taro festival.

These tribes were chosen for several reasons including the history of the tribe, affected by colonisation and the development of the mining sector, and the possibility of exploring the village's "memory of the water" through the "Vieux"[7] of the water clan. This access to the memory of the water and to the water clan is a key element in light of the strong oral tradition of the Kanak communities and the historical importance of these clans with regard to knowledge of this resource.

The empirical work called on several sociological and anthropological survey instruments. First, direct observation enabled us to see current practices first hand. This was followed by long periods of open interviews with the elders and the members of the water clan with a view to understanding and retracing the history of the conduits and how they were used. Some thirty interviews were conducted during the study, enabling us to analyse the factors underlying the disappearance of the taro plantations and shed light on current water practices and uses. These interviews were complemented by other interviews conducted in the town halls, which provided an insight into the organisation of the current drinking water network as well as the role of the municipalities and their link to the tribe. This field work was completed by a study of the archive provided by the SANC (New Caledonia Archives Bureau, Anon, n.d.-a, b, c, d, e, f, g), which helped complete and date elements identified in the surveys. These documents primarily concerned complaints relating to pollution and negotiations aimed at implementing new networks or catchments.

6.3 Results: From the Management of Open Irrigation Channels to Underground Drinking Water Networks

The results are presented in two stages. We initially reconstructed the operations of the taro plantations in the 1950s as described by the "Vieux" and members of the water clan. Based on local oral history and on the memories of the elders, these

---

[6] The former mining dumps upstream from the catchment areas contributed to silting up the rivers, and are still doing so, to such an extent that certain watercourses are officially qualified as silted or over-silted rivers. This silting up changes the morphology of the rivers, characterised by the river beds becoming wider and shallower (Garcin et al., 2017), and alters the inhabitants' practices and their relationship with the rivers (Gosset et al., 2019).

[7] Locally, the terms "Vieux" and "Vieilles" are a respectful means of referring to the older inhabitants.

operations, described as "traditional" by those interviewed, represent the starting point for the historical reconstitution presented in the second part of the results, but is itself already a system that has been transformed, in particular by colonisation. Using archives and interviews, we then reconstructed the history of forms of water management, focusing on the irrigated taro plantations, across three major periods: colonisation, the development of the mining sector and the development of drinking water networks.

6.3.1 "Traditional" Management of the Taro Plantations Shows a Collective Spatialised Organisation Built Around the Rivers and Channels

The organisation of the tribe around the river and the irrigated taro plantations was reconstructed thanks to the water clan elders. In the Nékliaï tribe, the water master drew the organisation of the taro plantations using his childhood memories and the vestiges he had explored during his horticultural and hunting activities. His drawing, which was subsequently digitised (Fig. 6.2), shows how the taro plantations functioned during the 1950s and represents what the elders see as the "traditional" operation of the taro plantations.

The tribes were organised around the river and the open water conduits. Water management felt to the water master, who was a member of the water clan. He maintained the springs and managed the division and distribution of the water among the clans. Like other traditional know-how, the traditional techniques associated with sharing and managing the water were linked to spiritual gifts, transmitted by the elders (ADCK, 2017). These beliefs are the very foundation of Kanak customs which govern both Kanak society and natural resources, in particular water (Sand, 2016).

The clans were spread across both banks of the river. Each clan benefited from a part of the river near their dwelling and was responsible for its upkeep. Everyone had access to large, very deep water holes used for fishing, washing or drinking water. "*At that time, we children had a little bucket to go and fetch water from the river. We used that water to wash, cook and drink.*"(testimony of a "Vieux" in the Nekliaï tribe).

A "pantry" was also marked out in the river for a period of four to five years. This pantry (a large water hole) constituted a reserve and a store of fish for major customary occasions such as deaths and weddings.

A network of open channels irrigated the yam fields, the taro and the gardens. In Nékliaï, the taro plantations were fed by two water conduits (conduits A and B in Fig. 6.2). Each conduit had its own spring located in the mountains of the Boulinda Massif. They ran down towards the tribe, feeding the taro plantations of the different clans. Each taro plantation was irrigated by small channels connected to the main channel of the conduit.

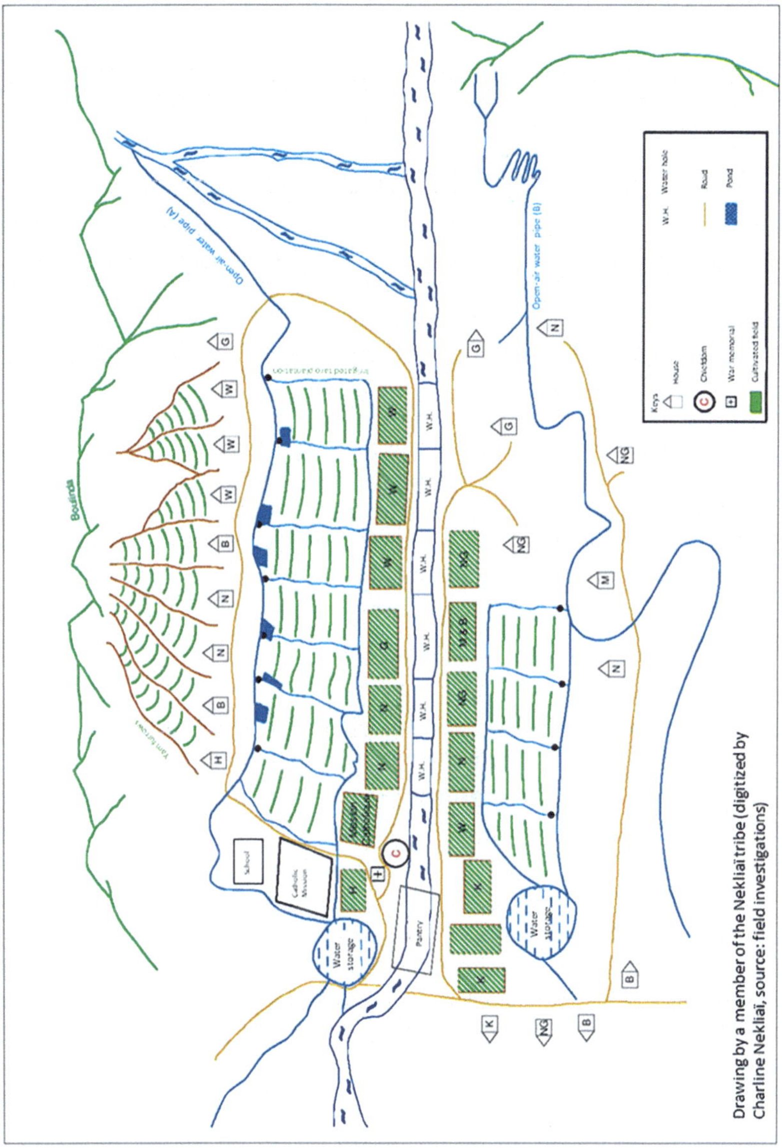

Fig. 6.2 Social and spatial organisation of the Nekliaï tribe around the taro plantations. Drawing by a member of the Nekliaï tribe, digitised by C. Nekiriai (source: surveys, Nekiriai, 2017)

To share and adjust the flow of water between each taro plantation, the water clan placed stones in the conduit to direct the water both towards the crops and in the main channel. Furthermore, at the source of the two water conduits A and B, there was also a conch, a sacred marine shell referred to locally as a "*Toutoute*". The water masters used this shell to "*control the water at its source*". Only the water clan possessed this know-how.

All the members of the tribe were required to help maintain the main channel. On the advice of the water master, the tribe chief decided when the channel should be cleaned and summoned the entire tribe to perform this task.

Each clan had its own yam field behind its dwelling on the valley floor. The yam fields were organised so that neither rainwater nor irrigation water would stagnate between the furrows. On each side of these fields, gutters were dug in the ground in order to evacuate the surplus rainwater towards the main water conduit.

In Canala, the taro plantation irrigation network was different, based on a single source and a much longer open conduit supplying several tribes. The organisation of each tribe around the main conduit and secondary conduits was nevertheless very similar to that observed in Nékliaï.

The sophistication of this format highlights the collective and spatial organisation of the tribe around water. This organisation, which lasted until the 1950s, reflects an intensive use of space and resources, also noticed by certain archaeologists (Sand, 2016), based on a collective and hierarchical management of water use, be it for agricultural or household purposes. As shown by the systems described by Hunt (1974) and, more recently, by Aubriot (2004), this management of taro plantation irrigation is entrenched in a set of family relationships and beliefs.

6.3.2 From Open Channels to Underground Drinking Water Networks: A Historical Retrospective of Technical and Social Transformations

Based on this description of the "traditional" functioning of the taro plantations, we reconstructed the history of water management methods in the tribes studied across three main periods: (i) a period prior to the 1950s, strongly impacted by colonisation, during which the system functioned differently form the situation described from living memory, (ii) a period impacted by the development of the mining sector (from the 1950s), and, more recently, (iii) a period impacted by the development of drinking water networks.

6.3.2.1 Colonisation and "Simplification" of Know-How

The way in which taro plantations were managed, as described by the "Vieux" in Fig. 6.2, was simply one step in the development of water management methods. Before the 1950s, colonisation and the decision to locate Kanaks in reserves led the tribes to adapt their lifestyle and revise their organisation of the taro plantations.

Figure 6.3 summarises the main impacts of colonisation on tribal water management, as discussed during the different interviews and analysed in light of the archives and literature.

Archaeologists have demonstrated the existence of major engineering works and large-scale taro plantations (some of the largest in the Pacific) dating back to before the arrival of the colonists. Although we do not have the necessary elements to reconstruct the operations of these major works, some vestiges can still be seen in Grande Terre (Dotte-Sarout et al., 2013). The scale of these works suggests that a high level of mobility and discussion was necessary between the clans on a large scale in order to ensure joint water management and distribution.

Three key events were identified in the bibliography and mentioned during the field surveys as having gradually dismantled this large scale organisation:

- first, the drastic reductions in the population associated with illness introduced by the first explorers (starting with Cook in 1774) led to changes in customary organisation (Sand, 2006). As elsewhere in the Pacific, these population decreases doubtless altered the customary spatial process (Dotte-Sarout et al., 2013);
- while no publication refers to knowledge pertaining to water, the arrival of missionaries in the 1850s disrupted traditional know-how and rites (Bensa & Leblic, 2000; Demmer, 2002). The arrival of the missionaries is still present in the collective memory of the tribes studied. "*My mother's mother said…: when the missionaries arrived, they told all the grandmothers and grandfathers to give up their baskets.*[8] *There is a place up there where they had to take them*" (comments made by the Nékliaï tribe);

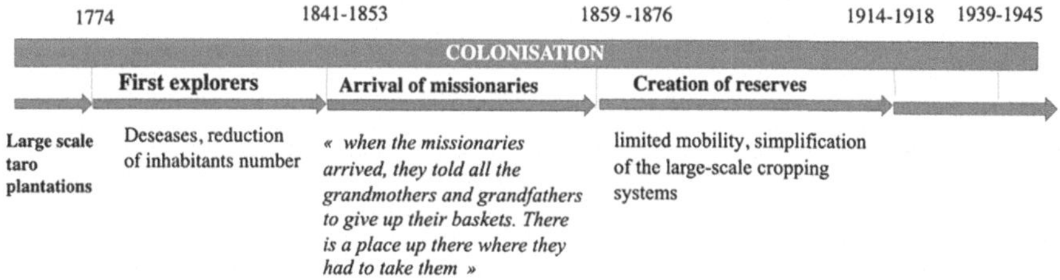

Fig. 6.3 Impacts of colonisation on the management of the taro plantations

[8] In these baskets, the Kanak place ingredients (plants) and cult objects/fetishes (stones, wood representing ancestors) to practise Kanak medicine, which corresponds to the animistic beliefs held by the Kanak prior to the arrival of the colonists. The techniques used to control water are part of this medicine.

- finally, the creation of reserves is clearly quoted as an element which altered the large-scale organisation. This is illustrated by the comments of one "Vieux" in the Nékliaï tribe: "*They rounded the tribe up there in that little area. We need to organise ourselves to produce enough for everyone and share the resources.*" These reserves limited mobility and discussions about water between the clans and meant a new organisation had to be developed within the restricted area of the reserves. According to Haudricourt (1964), the creation of reserves also led to a simplification of the large-scale cropping systems, in particular to the disappearance of the large yam ridges and the introduction of new, hardier varieties.

Colonisation, and more particularly the creation of reserves, was thus the underlying factor of the initial changes in the spatial and social organisation of the tribes. These changes were accompanied by a simplification of traditional know-how, especially with regard to water and cropping systems. The "traditional" functioning described in Fig. 6.2 is thus already the result of several phases of adaptation linked to colonisation.

6.3.2.2 Development of the Mining Sector and Replacement of Open Channels by Underground Channels

According to the surveys, the development of the mining sector from the 1950s onwards significantly changed the methods of managing the taro plantations described in Figs. 6.2 and 6.4.

First, "*the men in the tribe went to work in the mines*". This was the first phase in the decline of the taro plantations. The departure of the men to work in the nearby mines caused a twofold loss of knowledge and labour, with the men no longer present to manage and maintain the conduits.

In Nékliaï, the women married to the water clan took the decision to maintain the conduits without the consent of the men. "*Thinking they were doing the right thing, they cleared the channel with spades and widened it. If it is widened it, the current is reduced*" (a "Vieille" in the Nékliaï tribe). In the memories of the tribe, this event marked the early dysfunction of the taro plantations associated with the development of mining activities.

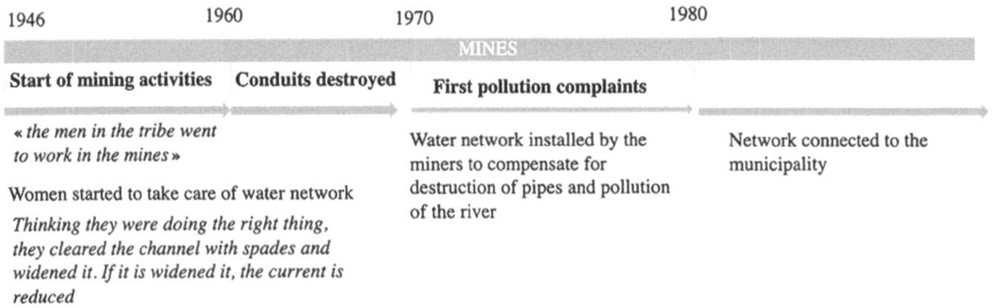

Fig. 6.4 Impact of the development of mining activities on the Nékliaï tribe

At the same time, several mines began operations in the area around the tribe. In Nékliaï, water conduit B was destroyed in 1964 by a colonist's mine located to the east of the Boulinda Massif on the left bank of the river. "*He came with a bulldozer then destroyed the conduit*" (comments of a clan chief in Nékliaï). This marked the end of the taro plantations on the river's left bank (cf. Fig. 6.2). The development of another mine upstream also created pollution in Nékliaï, in particular through the dust created by the trucks as well as the unusable waste and residue from the nickel mine dumped in the river. As the archive documents show, several complaints were filed with the mining authorities from the 1970s both by the tribes and the farmers in the region.

The complaints associated with the pollution gave rise to discussions within the tribe and with the miners. According to the people we spoke to, the customary and mining authorities came to an oral agreement that if the mining operators polluted the water in the river and the conduits, they would have to supply the tribe with drinking water by installing reservoirs, underground conduits and taps free of charge. Around 1970, the miners thus financed and installed the first underground conduits to offset the pollution. The underground conduits were installed under the open irrigation channels. Both the water and the maintenance of the conduits were free.

6.3.2.3 Drinking Water Networks and the Role of the Municipality

The third main phase of change in management mentioned by the elders in the tribe relates to the arrival of the drinking water networks (Fig. 6.5).

From the 1970s onwards, the installation of the first underground pipes followed by drinking water networks changed tribal lifestyle. With taps, the inhabitants no longer needed to go to the river or water points to obtain drinking water. Furthermore, they became less reliant on fishing as they began buying food from shops and purchased fridges. The river and water points were longer seen as pantries or water stores and maintained to a smaller extent. "*With the taps, the tribe turned its back on the river*" (comment made in the Mîa tribe). We can therefore observe a gradual change in the spatial and social organisation of the tribes, which were no longer organised around the river and the water points but around the water and electricity networks.

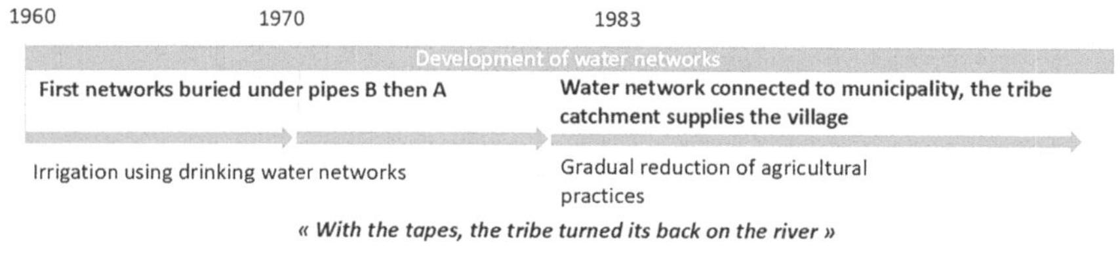

Fig. 6.5 Development of drinking water networks and the end of traditional management

In the 1980s, the municipality took responsibility for maintaining and managing all the water networks. This communalisation of the networks and other infrastructures implemented by the miners can be observed in several communities at this time (Le Meur, 2009). Following the closure of the nickel mines in 1983 in Nékliaï, the drinking water networks built by the miners were transferred to the municipality. The Nekliaï catchment was then connected to the municipal network to supply not only the tribe but also a part of the municipality (SANC documents). Accordingly, municipal management gradually replaced traditional management around the taro plantations. While resource management on customary land remains under the aegis of the tribes from a purely legal point of view (Massenavette, 2011), the municipality now plays a key role in the distribution of drinking water on customary land. It is responsible for supplying the population with drinking water (the mayor is responsible for public hygiene) and ensuring the upkeep of the water networks (a task it can delegate). Whenever it needs to invest in a water network construction project, it receives and manages subsidies from development contracts or other types of investment fund.

At the same time, certain fields continue to be irrigated and gardens watered using water from the underground conduits underneath the old open channels. With these open channels replaced by drinking water networks, drinking water is predominantly used for irrigation and watering.

6.4 Discussion: Shedding Light on Social Practices and Transformations and Creating the Link

6.4.1 Focusing on the Networks and the Associated Changes Sheds Light on the Social Transformations at Work in the Tribes: Erosion of Knowledge and Social Reorganisation

The study highlights the gradual impact first of colonisation then of the development of mining on customary organisation relating to water. Replacing irrigation channels by drinking water networks resulted in a transformation in the spatial organisation of the tribes, which no longer focused on the river and open channels but on access to underground water networks. The links to river were gradually eroded. With the arrival of taps, the tribe "*neglected the river*". Several practices linked to management of the river have tended to disappear, including fishing for major events, maintenance of the springs by the water master or the upkeep of the water holes and banks by the clans living near the river. Today, this upkeep is less and less the task of the inhabitants. From a legal standpoint, the municipality is not responsible for this upkeep and the river's silting is exacerbated by the lack of maintenance.

The analysis also highlights the gradual loss of traditional knowledge and rites. An elder in Nékliaï said that "*By placing the channels underground, we have lost our knowledge and forgotten our links.*" While the customary organisation of the tribes has changed with the development of the water systems (the water master no longer plays a leading role in sharing water), certain rules have nevertheless endured. Through its totem, the water clan is still clearly defined (the red eel in Nékliaï, for example). This clan is still the keeper of the water and protector of the springs. It is still often consulted with regard to tribal water management issues and is also the primary and preferred entrance point when understanding the history of water in the tribe.

In a society where the distribution of and means of handing down knowledge are pivotal (Bensa & Leblic, 2000), focusing on the water networks helps highlight social transformations. As emphasised by other authors, technical objects shed light on social dynamics (Fofack, 2018). In the Kanak tribes, this can be seen by analysing the networks that reflect the social structure linked to knowledge and of the "network of links" with the tribe.

6.4.2 Analysing Technical Developments with Regard to the Hydro-social Cycle Sheds Light on Current Water-Related Practices and Uses

Focusing on water channels with regard to the hydro-social cycle, including the cultural and social aspects of water, highlights current tensions linked to municipal management. This is illustrated by two examples in particular: the system for billing water and water wastage.

6.4.2.1 Water "Wastage"

Water consumption in New Caledonia is often very high and can reach several cubic metres per inhabitant per day within the tribe (Davar, 2018). Even without precise figures for the tribes in Nékliaï and Mîa, tribal water consumption is often deemed to be higher than in the village. "*In the tribe, people leave taps open*" (interview at Poya town hall); "*It is need to water the taro*" (interview at Canala town hall). In both tribes, the irrigation channels have disappeared to be replaced by underground pipes used for drinking water and managed by the municipalities. In Nékliaï, the drinking water networks were built under the former water channels traditionally used for irrigation purposes. Many inhabitants use tap water to water their garden continuously, as they did with the open water channels. These watering and irrigation practices have remained firmly entrenched in people's daily lives. Furthermore, for the inhabitants of the tribe, the notion of "wastage" has no meaning; "*it is not wastage, the plants need water and the water returns to nature*" (interview in the Mîa tribe). Water circulates, creating the link between the elements of nature.

The analysis focusing on the history of the networks, including the cultural and symbolic aspects of water, provides a better understanding of the different ways tribes uses water. These uses are linked to habits and the symbolism of water, an element which must circulate, and not to a technical rationale aimed at saving water. The analysis highlights a conflict of representations between the municipalities and the clans, a contrast visible in the prevailing discourse of the water services administrators which opposes the "rationality" of the municipalities and the "irrationality" of the clans. From an operational standpoint, this new vision of "wastage" can promote increased awareness of water savings made by the municipalities among users.

6.4.2.2 The Issue of Paying for Water

Most municipalities in New Caledonia are currently encountering difficulties linked to the non-payment of bills. In certain municipalities, water is free and in those which apply a billing system, payment delinquency rates can be as high as 80% (Davar, 2018). As an elected representative of the municipality of Poya explained, "*the position adopted in the tribes is "my water, my water*". Even if the municipality has managed water maintenance and treatment for over 30 years, the idea of paying for water often remains a source of high tensions with the tribes. The analysis helps understand the source of these tensions. On the one hand, the springs and catchments of the municipal networks are, for the most part, located on customary land (this is the case in the areas studied and for 52% of catchments in New Caledonia). On the other hand, the first networks were, in some cases such as in Nékliaï, obtained as compensation for pollution in accordance with oral (and thus often little known) agreements on the free provision of water. The analysis identifies this history, handed down orally, and provides a better understanding of the origin of this rhetoric. It also highlights the discrepancy described by other authors (Trépied, 2004, 2011) between "*customary sovereignty*" (over water) and the municipal institutions. As numerous authors have shown (for example Obertreis et al., 2016), the technical choices linked to water management give rise to a transfer of responsibilities, competences and knowledge, thereby altering the balance of power relating to water over time.

6.4.3 Transfer of Responsibilities and Refocusing of Knowledge and Power

The analysis shows that municipal management has gradually replaced "traditional" management of the taro plantations. The erosion of traditional knowledge has resulted in a transfer of responsibilities in the field of water management. The municipality now plays a key role in distributing drinking water on customary lands even though, from a purely legal standpoint, this does not fall under its auspices. Extremely costly in light of the large size of the municipalities[9] and the low population density, the implementation and development of water networks means that the municipalities have to implement a pricing system that ensures balanced budgets. Implementation of this pricing system is all the more complicated as it is imposed by the municipality and the government. Some people feel that this pricing system symbolises a form of appropriation of the resource by the municipality.

The study thus reveals a fear of appropriation of the resource both via the water works (the networks) and via the appropriation of management responsibilities either by the municipalities or by the New Caledonian government and the French state, who can impose rules for using the resource or the pricing system necessary to balance the municipalities' budgets. In this case, it is not the content of the knowledge associated with the technology which crystallises the tensions, but the actors who possess the knowledge and their link to power. As explained by Mahias (2011), "*the definition and recognition of technical knowledge, and more particularly its appropriation and its transmission, are always at the heart of social relations. They constitute instruments of power and socio-political issues, and are always subject to social control.*"

In New Caledonia, these tensions and fears of appropriation are extremely sensitive, both in the current political context of negotiated decolonisation and in light of the strong links between land issues and water issues. The conflicts relating to land and those relating to water are closely linked. Due to the particular status of customary lands, the municipalities call on the customary authorities to implement water and sanitation projects. To draw up a new water development plan, they create a link with the clan and district councils in order to limit land issues. Implementing or renovating a water network can nevertheless be a very long process, "*sometimes it is about a coconut palm or a mango tree, but for the person who is there, it is a reference point, the tree that their grandfather planted... it is not sentimental, it is a reference asserting their belonging to the land*" (interview with an elected representative of the municipality of Poya). Beyond the appropriation of the resource, the implementation of new water networks is often associated with a fear of land appropriation.

The analysis thus highlights the fears linked to the appropriation of knowledge and resources. With technological developments and the transfer of water management competences, new balances of power are established which add complexity to the analysis as many tribal representatives are also elected representatives in the municipalities or in government. Management of the underground networks should therefore result in new links and new knowledge, including within the tribes.

[9] The New Caledonian municipalities are particularly large, covering an average of 562 km^2 which corresponds to four average cantons in France.

6.4.4 Recreating the Link by Deciphering the History of the Channels?

Throughout the surveys, the channels—the manifestation of water and the link—became a preferred means of expressing the tensions, adaptations and changes at work. Understanding the networks and reconstructing them based on the stakeholders' words have helped portray an oral history; changes, be they imposed (colonisation) or chosen (work in the mines); deviations from customs or hierarchies between clans; or non-compliance with symbolic or cultural references often hidden since the period of religious colonisation. Addressing this via the technical aspect of the taro plantations enabled us, in part, to tap into the "water memory" through the water clans. The history of the pipes served to help navigate between the "world of the tribes" and "the world of the municipality and the government", as well as within the tribes themselves. The testimonies collected rarely referred to past conflicts relating to management of the taro plantations, despite the questions asked on this subject. Necessarily selective, the memory of the elders focussed on highlighting elements of the "proper" functioning of the taro plantations in compliance with the local social organisation. As demonstrated in other fields (Riaux et al., 2015), oral memory revises history to retain significant signs in relation to current issues, and society understands its system through elements that it finds significant. In our case, oral memory and historic reconstitution illustrated the coherence of the remains of the taro plantations which were in such a condition that it would have been difficult to understand how they worked. This enhanced visibility enabled the stakeholders to express their perceptions of contemporary interventions in and modifications to water management while legitimising a local history.

With this in mind, following the development of the channels creates and recreates the link between the tribes and the institutional actors as well as within the tribes themselves.

6.5 Conclusion

In New Caledonia, a country where water symbolises "the link" between men and with nature, analysing the hydro-social cycle of water through the water networks appears to be a self-evident approach to understanding social transformations. Focusing on technical objects, including their historic and cultural dimension, facilitates discussion and debate in the tribes. Social changes appear through technical change and may prove delicate to illustrate and express in the field in light of the political context marked by a dynamic of decolonisation under negotiation. Addressing the theme through technical objects reveals major transformations in knowledge and power relating to water. The analysis therefore sheds light on the impact first of colonisation then of mining sector development on customary management and organisation. It highlights certain changes in the spatial and social organisation of the tribes as well as the fears concerning the appropriation of resources.

Following the first referendum concerning full sovereignty and with the launch of a new policy relating to water (Gouvernement de la Nouvelle-Calédonie, 2019), the analysis examines the policies at play with regard to the complex relations that could be formed between knowledge, technology and power regarding access to and distribution of water. The implementation of the new water policy was led by the government's youngest two ministers, one a Caldoche and the other a Kanak. As the political discourse associated with the implementation of this policy suggests, water could become a symbol of entente, of the "link" to protect a common good.

References

ADCK (Agence de développement de la culture kanak) (2017). *La culture kanak, une identité affirmée, association Wasapa art kanak*. Disponible sur: wasapaartkanak.free.fr/archives/LaCultureKanak.pdf

Anon (n.d.-a). *Documents d'archives—SANC* (Service des Archives de la Nouvelle-Calédonie).

Anon (n.d.-b). *Boîte 339W-172 Affaires communales de Poya, Nékliaï-Népou.*

Anon (n.d.-c). *Procès verbal de la réunion du 20 mars 1970 qui porte sur un différend entre les éleveurs et les miniers au sujet de la pollution et de la qualité de l'eau et illustrant la résolution des différends avec les populations locales par compensation financière.*

Anon (n.d.-d). *Boîte 339W-160—Affaires communales de Poya: Compte rendu de réunion du comité de l'environnement du 22 aout 1972, Étude à effectuer et conditions à remplir par les exploitants miniers, détails donnés par zone d'exploitations.*

Anon (n.d.-e). *Correspondances de 1975 entre le maire de Poya et le président de la commission des pollutions concernant les décharges minières sur la commune.*

Anon (n.d.-f). *Boîte 339W-150—Affaires communales de Poya.*

Anon (n.d.-g). *Dossier complet sur l'adduction d'eau de la commune de Poya 1983–1984.*

Aubriot. (2004). *Leau, miroir d'une société: Irrigation paysanne au Népal central*. CNRS. 321 p.

Aubriot, O., & Riaux, J. (2013). Savoirs sur l'eau : les techniques à l'appui des relations de pouvoir? *Autrepart, 65*, 3–26.

Barbier, B., & Trépos, J.-Y. (2007). Humains et non-humains: Un bilan d'étape de la sociologie des collectifs. *Revue d'anthropologie des connaissances, 1*(1).

Bédoucha, G. (1987). *« L'Eau, amie du puissant » : une communauté oasienne du Sud tunisien.* Éditions des archives contemporaines. 428 p.

Bensa, A., & Leblic, I. (2000). *En Pays kanak: Éditions de la Maison des sciences de l'homme, coll.* Ethnologie de la France. Cahiers 14.

Budds, J., & Linton, J. (2014). The hydrosocial cycle: Defining and mobilizing a relational-dialectical approach to water. *Geoforum, 57*, 170–180.

Davar (Direction des affaires vétérinaires, alimentaires et rurales) (2018). *Forum H2O: quelques chiffres clés, document présenté le 18,19 et 20 avril 2008 à Nouméa*, disponible en ligne sur www.forumh2o.nc

Demmer, C. (2002). *Les héritiers d'Éloi Machoro (1941–1985): une génération nationaliste au pouvoir à Amâà et Kûöö, villages de Xârâcùù (Canala) Nouvelle-Calédonie, thèse de doctorat: anthropologie sociale et ethnologie*. EHESS. 565 p.

Dotte-Sarout, E., Sand, C., Bole, J., & Ouetcho, A. (2013). Archéogéographie et processus de territorialisation pré-contacts au nord de la Grande Terre (Nouvelle-Calédonie). *Journal de la Société des Océanistes, 136137*(1), 195–207.

Fofack, R. (2018). *La société des eaux cachées du Saïss. L'analyse d'un basculement autour de l'extraction des eaux souterraines profondes au Maroc.* Diss. thèse de doctorat, sociologie, Université Paris Nanterre.
Garcin, M., Richard, D., Liébault, F., Recking, A., Piton, G., Sabinot, C., Worliczek, E., Lesimple, S., Bertrand, M., Gastaldi, Y., Burlat, T., & Gosset, L. (2017). *Gestion du passif minier—Guide méthodologique.* CNRT Nickel & Son Environnement. 82 p., 20 fig., 15 tab.
Gosset, L., Sabinot, C., & Worliczek, E. (2019). Quand cyclones, pluies et pollution interrogent les liens des Kanak à leurs rivières et participent au renouvellement des savoirs écologiques (Thio, Nouvelle-Calédonie), Ethnographiques.org. *Numéro spécial Approche anthropologique des changements climatiques et météorologiques.*
Gouvernement de la Nouvelle-Calédonie (2019). *Politique de l'eau partagée, schéma d'orientation pour une politique de l'eau partagée de la Nouvelle-Calédonie.* https://davar.gouv.nc/sites/default/files/atoms/files/pep_-_schema_dorientation_pour_une_politique_de_leau_partagee_de_la_nouvelle-caledonie.pdf
Haudricourt, A.-G. (1964). Nature et culture dans la civilisation de l'igname: l'origine des clones et des clans. *L'Homme, 4*(1), 93–104.
Hidalgo, J. P., Boelens, R., & Vos, J. (2017). De-colonizing water. Dispossession, water insecurity, and Indigenous claims for resources, authority and territory. *Water History, 9*(1), 69.
Hsiao, E. C. (2012). Whanganui River agreement: Indigenous rights and rights of nature. *Environmental Policy and Law, 42*(6), 371–375.
Hunt, E., & Hunt, R. (1974). Irrigation, conflict, and politics: A Mexican case. In T. Downing, McGibsong (dir.), *Irrigation's impact on society* (pp. 129–157). Tucson: University of Arizona Press.
Jackson, S., & Altman, J. (2009). Indigenous rights and water policy: Perspectives from tropical Northern Australia. *Australian Indigenous Law Review, 13*(1), 27–48.
Janty, G. (2013). Capacité d'adaptation des pratiques traditionnelles de gestion et de partage de l'eau dans l'oasis de Figuig (Maroc). *Autrepart, 65*, 3–26.
Le Meur, P.-Y. (2009). Opérateurs miniers, gouvernementalité et politique des ressources à Thio, Nouvelle-Calédonie. In *Pacific science intercongress proceedings* (5 p.).
Leblic, I. (2005). Pays, "surnature" et sites "sacrés" paicî à Ponérihouen (Nouvelle-Calédonie). *Journal de la société des océanistes, 120–121*, 95–111.
Lejars, C., Bouard, S., Le Meur, P.-Y., & Ferrand, N. (2016). Exploring water governance in Pouembout (New Caledonia). In *Water governance in Oceania workshop, 28–29 June 2016, Nadi, Fidji.*
Lemonnier, P. (1980). *Les salines de l'Ouest: Logique technique, logique sociale* (222 p.). Paris: Maison des sciences de l'homme/Presses universitaires de Lille.
Mahias, M. C. (éd.). (2011). Construire les savoirs dans l'action. Apprentissages et enjeux sociaux en Asie du Sud, Paris, Éditions de l'EHESS, coll. « Purusā rtha », 29, 2011, 293 p.
Massenavette, C. (2011). La reconnaissance législative du domaine fluvial de la Nouvelle-Calédonie: la fin d'une longue période d'insécurité juridique. *Revue juridique de l'environnement, 36*(1).
Nekiriai, C. (2017). *La gestion de l'eau sur terres coutumières en Nouvelle-Calédonie: Évolution des représentations, des valeurs et des pratiques* (70 p.). Mémoire de master de l'université de Nantes.
Obertreis, J., Moss, T., Mollinga, P., & Bichsel, C. (2016). Water, infrastructure and political rule: Introduction to the Special Issue. *Water Alternatives, 9*(2), 168–181.
Récalt, C., Rouvière, L., Mahdane, M., Errahj, M., & Ruf, T. (2013). Aménager l'espace, canaliser l'eau et orienter le pouvoir. Réflexion sur deux modèles inédits d'aménagements fonciers radioconcentriques en France et au Maroc. *Autrepart, 65*, 3–26.
Riaux, J., Giraldi, M., & Nouri, H. (2015). Quand l'histoire orale s'invite dans la gestion de l'eau. L'espace irrigué de Kerma (Tunisie). *Espaces et sociétés, 1–2*(160–161), 155–172.
Sabinot, C., & Lescureux, N. (2019). Local ecological knowledge and the viability of the relations with the environment. In O. Barrière, M. Behnassi, G. David, V. Douzal, M. Fargette, T. Libourel, M. Loireau, L. Pascal, C. Prost, V. Ravena-Cañete, F. Seyler, & S. Morand (Eds.),

Coviability of social and ecological systems: Reconnecting mankind to the biosphere in an era of global change (pp. 211–222). Springer.
Sand, C. (2006). Historical Archaeology in Island Melanesia: First Research on the Convict Settlements of New Caledonial. *New Zealand Journal of Archaeology, 27*, 145–161.
Sand, C. (2016). *La gestion des communs dans les sociétés anciennes du Pacifique*. Colloque sur les communs, novembre, Agence française de développement, Paris.
Seemann, M. (2016). Inclusive recognition politics and the struggle over hydrosocial territories in two Bolivian highland communities. *Water International, 41*(1), 157–172. https://doi.org/10.1080/02508060.2016.1108384
Sénat coutumier de la Nouvelle-Calédonie (2014). Charte du peuple kanak: Socle commun des valeurs et principes fondamentaux de la civilisation kanak (32 p.)
Swyngedouw, E. (2004). *Social power and the urbanization of water flows of power*. Oxford Geographical and Environmental Studies. 228 p. ISBN: 9780198233916.
Trépied, B. (2004). Coutumiers kanak contre élus FLNKS. La bataille de l'eau à Koné (Nouvelle-Calédonie, 2004). In B. Bosa & E. Wittersheim (Eds.), *Luttes autochtones, trajectoires postcoloniales (Amériques, Pacifique), Paris, Karthala* (pp. 27–56).
Trépied, B. (2011). Des conduites d'eau pour les tribus. Action municipale, colonisation et citoyenneté en Nouvelle-Calédonie. *Revue d'histoire moderne et contemporaine, 58*(4).
Wagner, J. R., & Jacka, J. K. (2018). *Island rivers: Fresh water and place in Oceania, Acton*. ACT ANU Press.
Wateau, F. (2002). *Partager l'eau. Irrigation et conflits au nord-ouest du Portugal* (277 p.). Paris: CNRS/Maison des sciences de l'homme.

Chapter 7
The Five Kanak Waters

Sonia Grochain

Abstract This chapter provides insight into the role of water among indigenous peoples through narratives and the history of property. This study of indigenous water takes into account a range of acts that bring tradition to the fore. Through deliberate actions, tradition becomes a consciously practised and preserved aspect among a significant number of the Kanak people. Today we see it in the form of grandparent-to-grandchild transmission, association, a variety of oral traditions, via cultural residences as well as in works devoted to tales, toponymy, legends, narratives and ritual stories. Through this narrative analysis five Kanak waters are distinctly identified. After presenting these five types of water, we discuss how they could be taken into account, or at least partially respected in recent water policy.

Keywords Indigenous waters · Narrative · Legend · Ritual

7.1 Introduction

The manner in which water is managed in New Caledonia differs depending on the land on which the resource is located, i.e. either on customary land belonging to the Kanak (Saussol, 1979; Herrenschmidt & Le Meur, 2016) or elsewhere. However, little is known about the characteristics of "customary land water", and the

---

This chapter draws inspiration from an article that was initially published in French in the review Geocarrefour. The review and the editors of the special issue have granted permission for a translation into English and publication in this book. **Sonia Grochain,** « Les cinq eaux des Kanaks. L'eau des clans, des terres coutumières, des vieux, l'eau pure et l'eau collective», Géocarrefour https://doi.org/10.4000/geocarrefour.20310.

S. Grochain (✉)
IAC (New Caledonian Institute of Agronomy), Nouméa, New Caledonia
e-mail: sonia.grochain@iac.nc

C. Lejars, S. Bouard (eds.), *Water Policy in New Caledonia*, Global Issues in Water Policy 32, https://doi.org/10.1007/978-3-031-88196-1_7

implementation of a new water policy needs to take into account Kanak identity when managing water resources.

The Kanak live in clans, or in families if the number of clan members has shrunk from generation to generation; there are 341 tribes who live on customary land. Their customs, i.e. the foundations of their social organisation, are based on the concept that the clan is made up of a group of lineages descended from a common ancestor. Each clan has a clan council made up of clan chiefs headed by a tribal chief and a clan council chair. The master of the land, the first occupant of the tribe, conserves the memory of the clan's or family's properties. They appoint representatives from the eight customary areas of New Caledonia - Djubea Kapone- Xaracuu-Ajie Aro-Paici Camuki-Hoot ma Whaap-Iaai-Drehu-Nengone, who are responsible for appointing the customary senators who sit on the Customary Senate, which forms an advisory body on Kanak identity for the New Caledonian government (see Chap. 1).

The memory of the five Kanak waters is recounted during ceremonial exchanges. The tradition that is passed on is not a fixed entity; it also evolves over time. Various leading figures, with complementary roles, work to ensure that the tradition continues, nurturing the present with the wisdom of the past and the vitality of the future. In addition to political figures such as Jean-Marie Tjibaou,[1] who was a fierce defender of his culture, there are those who collect or write down oral traditions, field leaders in each tribe, specialists in oral traditions, grandparents who pass on traditions to their grandchildren, customary senators, tribal chiefs and the chairs of clan councils. This elite group preserves the traditions that illustrate the history of tribal water.

This chapter studies indigenous water types and traditions, based on oral traditions and direct communication that has been recorded to preserve the traditions that describe the five waters of the Kanak (Grochain, 2020): clan water, customary land water, ancient water, pure water and collective water.

Water management on customary land figures among truly conscious traditions: "*intentionally created, transmitted by force, because they result from the necessities of living together. Traditions can be called conscious when they consist of a society's knowledge of itself and its more or less immediate past.*" (Mauss, 1969, p. 120). The French sociologist and anthropologist Marcel Mauss studied the traditions of indigenous peoples, from the Eskimos to the Aborigines, distinguishing between two types of tradition, pure social tradition and the tradition of natural and supernatural things, aiming to formulate the element of descriptive sociology. In this chapter, the tradition of natural and supernatural things (Mauss, 1969, p. 122) is studied using information on cosmography and beliefs related to water. Ultimately, understanding these traditions and perceptions of water raises questions about how to preserve water as a resource, as well as water as a cultural or mystical entity.

[1] Jean-Marie Tjibaou, 1936–1989. Political figure in Kanak nationalism, founder of FLNKS, assassinated.

7.2 Clan Water

Clan water comes from springs located on clan lands, as described in this oral tradition reported by Raymond Kona, a specialist in oral traditions; it is about the mountain of the Chéréy mê Nêpamwâdù clan:

> *Mount Pâbaii is said to be a standing man with his face turned towards Sarraméa and his back towards Canala. At its summit is a pond where rushes grow. This is one of the sacred sites where the Négropo River rises. On the western slope towards Sarraméa, there are the three thunder stones. According to legend, the first of these took in its mouth the "flèche faîtière*[2]*" of the great hut belonging to the Méyù of the Haouly tribe "Xêwiri", with the conchs hanging from it. The second and third are both called Sarèkârâmè, which means "to attack the eyes". It is forbidden to ask for directions at this particular spot, and one should walk with one's back to the two thunder stones so as not to see them, otherwise one could risk getting lost in the forest. At the bottom of Pââbaii towards Sarraméa lies the yam stone which belongs to the clan of our informant. During the period when the yams begin to grow, this rock takes on a green colour, which is the colour of the stalk, and when the yams are ripe, it takes on a white colour, which is the colour of a yam when it is about to be dug up. Above this stone grows a tuft of reed (xûja) used as a wrap for customary ceremonies (weddings, funerals, births)* (Kona Raymond, Xaracuu oral tradition, 2007, p. 14).

Maurice Leenhardt, a notable protestant ethnologist and missionary, has shown the importance of this clan-based division of the country, by describing 86 parcels of riparian land on the small Houaïlou River, and in particular by highlighting the role of the master of the land, Kavu, guardian of the family properties of his tribe. He is considered to be the first occupant of the tribe's home lands; he knows their entire history, as well as that of the tribal chief he welcomed onto his land.

> *To demonstrate the extent to which the use of the land is divided amongst the clans, here are the names of the various parties on the little Houaïlou River, and those who have fishing rights. The river is 50 kilometres in length and thus has 100 kilometres of banks. 86 parcels remain in the possession of an equal number of residents. Usage rights are passed on through marriage, war treaties, etc. and all these distinctions are recorded in the memory of the elders, which is the great book of the chiefs. We can consider that the country is inhabited and divided up in this way for hunting, fruit gathering and to benefit from irrigation* (Leenhardt, reprint 1980, p. 91).

7.2.1 Traditions, Totems and Gods

The tradition of natural and supernatural things is reported in several oral traditions and direct communications that the Kanaks have been passing down for two or three generations. The description of their clan water, the thunder or water stones needed for conveying water or for water and rain rituals according to the magic of meteorological phenomena are all part of the way the clan manages water (cf. Sarasin, 2009, p. 262). For instance, water stones are attached to the ends of bamboo pipes, thus allowing water to pass through them, which helps to keep the water clean. Thunder

[2] A carved roof top spear placed at the top of a hut.

stones are used for digging at a spring to make it gush forth, and water stones bring the water into motion.

Water belonging to the clan is also associated with Kanak beliefs, like those on the water totem, or with the clan's ancestral gods, who, according to the Kanaks, dwell in the clan's springs or waterways. According to the *Cahiers de 1918–1921* by Bwesou Eurjisi, the first Kanak writer to work with Maurice Leenhardt on the preservation of his own culture by drawing in these texts (Rognon, 2018), the water totem pole was installed when the Wêêxumé (a clan) began to found a tribe at Houaïlou (Aramiou & Euritein, 2003, p. 123). In the Ajie and Paici customary areas, there are around fifteen clans whose totem is water, but there is no need to have water as a totem in order to carry out traditional waterworks.

In these two customary areas, eight water-dwelling gods have been identified thanks to the research work of ethnologists Jean Guiart and Maurice Leenhardt (Guiart, 1963; Leenhardt, 1980). There would be many more if we were to refer to funerals: during these ceremonies, Kanaks always invoke the ancestor gods of each clan, representing their spring gods to the spiritual leaders of thunder and rain; one of their functions is to officiate for the return of the dead to their ancestors.

This knowledge of and belief in water has forged clan relationships between the great chiefdoms, just as much as it explains the disappearance of water when the clans quarrelled over the choice of religion, with one clan leaving with the water stone (Arthur Maramin, Xârâcùù oral tradition, 2011, p. 21 and Jean Paul Powi, Xârâcùù oral tradition, 2017, p. 59).

7.2.2 Sun, Rain and Thunder Stones

In addition to the water stone, others such as the rain, thunder and sun stones are described in the ritual accounts of water and rain, and have helped to establish the spiritual leader of thunder and rain in the ritual tradition. According to Bwesou Eurjisi, in the Ajie customary area, the rain and sun stones arrive via the boria (dance of the dead) of those who have died and who previously held these stones (Aramiou & Euritéin, 2003, p. 27).

For the French priest Pierre Lambert, who described Kanak culture at the end of the 19th century (Lambert, reprinted 1999, p. 297), the rain stone was put into pots with water prepared during a ritual to make it rain during a drought; it was also used at the request of tribal chief who wanted abundant showers from the sky. Francis Garnung mentions another rain stone loaded with cumulus clouds (Garnung, 2016, p. 15). Leenhardt described propitiatory rites for the cultivation of yams and taro (*Colocasia esculenta*) (1937, p. 66–68), which always associate water and rain with cosmography. At the end of these rites, young people today still choose an individual act of learning, such as cultivating arum, yams, water or traditional medicine. These rites are also used to ensure that the following day, after the hard work in the fields, the yams or taros will be fertile thanks to the light rain that falls on the shoots.

In 2002, Jean Marie Nêba, a specialist in oral traditions, brought back a ritual story about the "rain stone, thunder stone and sun stone". For him, it is traditional events and ritual stories that help to maintain this tradition through formulas and secrets to obtain rain, water and the return of the sun.

> *The spiritual leader for thunder (...), he knows the ritual for the three stones. That's what he told us. Roughly ten young shoots of xûja reed are harvested, taking care to leave the last two leaves and the central shoot. The spiritual leader goes to the site of the thunder stone. Facing it at an angle, he steps forward and throws the reed shoots, invoking the stone to gather the clouds in the sky so that the rain will fall. Once this ritual has been performed, the spiritual leader officiates in turn for the rain stone; the rain stone is the wife of the sun stone. The spiritual leader takes the köpu vine, which grows in the taro fields and is given to the children to drink. He wraps it around the head of the stone. He takes the stone and places it in a hollowed-out piece of houp (a tree called xu). Water is poured in to halfway up the stone. The water must come from a lake located in the dwelling of the lizard 'Kabaé', at a place called 'Sei-nô' in the commune of Kouaoua. An eel (pârâ), no larger than the fingers of a hand, is caught using a piece of wood and pricked, which, as it moves, causes water to splash onto the head of the stone. This is reminiscent of floodwaters that jump over rocks in rivers. As he performs these gestures, the spiritual leader invokes the spirit of the stone to make the rain fall. Once it has rained, the spirit of the sun stone must be invoked again to bring back the fine weather. The spiritual leader takes the sun stone, places it in a dry place and invokes the spirits. He places two pieces of dry wood next to the stone to smoke it and invokes the spirits for sunshine. To carry out these rituals, the officiant must observe strict rules concerning fasting, the consumption of food and medicinal plants. These measures are absolutely essential if the officiant does not want the spirits to turn against him* (Jean Marie Nêba from the Nûî clan, Xârâcùù oral tradition, 2002, p. 34).

Alongside these stones, the water stone ensures that water is conveyed in the bamboo to bring the water to the tap, and a water song tells the story. Using notched sticks, vines and plants, this song recounts the various steps for channelling water according to tradition to irrigate the large terraced taro fields (Leenhardt, 1932, p. 326–333).

7.3 Pure Water

Pure water, or "real water", is that which is stored in natural cisterns or wells, and which women boil for their daily consumption. The interest shown in this water by the Kanaks stemmed both from its quality, as it was considered purer, clearer and cleaner than other waters, and above all from its accessibility, as it was not allocated as property, and sometimes it was on the land of other clans.

> *In other places there are wells, beautiful water at old Groaiu's, next door at Apiazari's, over here there's a well. People used to get their supply there, the water was very clear and very clean, they said it was thanks to the elders, but the well was already there, it dates back a hundred years, a well one and a half metres in diameter. As for the water, the women used to cross over to fetch water from the wells of the Gopoéa (neighbouring clans) in the 1970's* (Grochain Clément, Chair of the Customary Senate, 2018).

The use of pure water has been preserved by a water tradition, especially in the Xârâcùù customary area and the three customary areas of the Loyalty Islands (Iaai,

Drehu, Nengoné); it is based on memorised tales that are handed down by grandparents to grandchildren. Today, three lessons are used to teach the water tradition:

1. making pure water, water flavoured with papaya and water naturally sweetened with coconut;
2. bodily positions and techniques in relation to water;
3. know-how to create irrigated fields for arum and *colocasia esculenta*.

Pure water was made using leaves from the niaoulis tree (Melaleuca) mixed with boiling water. The aim was also to make the water a part of one's life by caring for it with a bag made from the bark of the niaoulis tree. The tradition of making pure water was passed on in the company of the children of the clan, household or tribe. It is accompanied by the preparation of water flavoured with papaya and naturally sweet coconut water (Fig. 7.1).

The bodily positions and techniques in relation to water are also studied, especially when natural cisterns or water stones are used to build up the mass of the river during the construction of dams. These bodily techniques are also used by everyone, from childhood onwards, to put their thoughts in order. Maurice Leenhardt also describes the body of water as part of a set of Kanak body techniques: "Karo rhë the body of water that is the mass of the river" (Leenhardt, 1947, p. 68). The mass of the river is an opportunity to learn physical techniques, swimming and knowledge of water distribution. Water must also be controlled; one must learn to bend it for it to calm down and have every possible understanding of the real water in the flow of the river; the bravest young people will know how to carry the water. Avoiding drowning also means folding as much of the water as possible so that it moves away from the presence of the child who might find himself alone at the river. In this way, you can tell for sure whether a child is safe at the river and whether they will become a member of the indigenous society with their body techniques or practices related to water, as well as their knowledge of water conveyance. This knowledge is also acquired over time, often through community events organised by the young girls known as the mistresses of lightning, rain and thunder. It is also important to be able to identify the thunder stones and recall the locations, the places to dig for the clan's water source.

The third lesson for maintaining the water tradition consists of the irrigated fields for *Colocasia esculenta* (better known as taro plantations[3]) on a water channel, which is revealed by the ancestors and the gods. Two videos, one on the ten years of the irrigated taro plantations of the Mia tribe and the other on the irrigated taro plantation of the Emma tribe, are still used for instruction on how to make them in Xârâcùù (Ngoni, 2016; Moasadi & Diapoiri, 2008). Young girls and boys learn how to grow water taro on terraces, from preparation to planting, including irrigation, which is important because taro needs water (it must circulate properly on the terrace without stagnating). Their organisation, managed by the chair of the Mia clan council, has identified two hundred varieties of water taro, and promotes the

[3] See Chap. 6.

Fig. 7.1 Women drinking water. Source: ANC Photo by Serge Kakou

cultivation of water taro handed down by the ancestors, as well as the consumption of local products for a healthy diet. There are three types of water taro: Chef taro, Wako taro and Wé taro. The magic sun technique, which consists of reversing the trajectory of the sun's rays in the irrigation basins to reflect the sky, is also used in the fields in the mountains and plains to retain celestial events, particularly when organising the area for sowing. A third video, made on the island of Iaai, shows how to plant water taro with freshwater duckweed. On the Loyalty Islands (Iaai, Drehu and Nengoné), water is only present in the form of groundwater, the surface of which is covered by duckweed. In the Uakatr tribe, thanks to a permanent layer of freshwater duckweed on the water table, the cultivated area can be irrigated in all seasons, particularly when the weather is hot (Tjibaou et al., video and Iaai oral

Fig. 7.2 The life of the indigenous people. ANC collection Serge Kakou, album J.O. Haas, 3, 148FI 15-39

tradition, 2007). These three videos were produced between 2007 and 2016 as part of the program to safeguard tangible and intangible Kanak culture, by Emmanuel Tjibaou, director of the Jean Marie Tjibaou cultural centre (Fig. 7.2).

7.4 Ancient Water

7.4.1 Ancient Water from Before the Neolithic Period

Ancient water is mainly described by specialists in oral traditions, masters of the land who use it for the yam tradition, and young people who describe the ancient water and the history of a clan. The Department of Infrastructure and Territories has recorded requests from elders for the purchase of land and the water on it, as well as for the organisation of bequests of land and water exclusively to elders (over the age of 50 years), after the death of all the purchasers. Ancient water is therefore located on the property of the elders. The oral tradition on the history of the Xwâkwé clan, written by Noël Gouémoin, a specialist in oral traditions, provides more detailed information on the events that took place there, the use of the ancient water source by travellers that has given its name to a large clan:

> *The Xwâkwé clan was based on Wânyîî territory (La Ouenghi) in the Nèbwèi habitat. They headed for the Mirînyâ mê Töxwâgwè pass. They came and sat down at Xwa Xwâkwé. It's a name given to a place by the elders because they own it. One of the elders told them to make a place to live on the outskirts. They sat down at this place and the path to draw water was far away. They went back down to draw water until they reached the dwellings, then they went back up again. They began to say:*
>
> *'Elders, I say we sit here, and the way to fetch water is too long. We got tired fetching water'. They lived in a hut. That one, the elder heard them and took pity on them. He did the sanctification. Those there, those who remained with him did not know what this elder was doing. He asks for the sanctification of this water. In this house, he scatters with straw. They slept that night, then felt a chill in their hut. They wondered what was wet under the straw. It was this water that the elder was looking for sanctification. This water was coming out of the house where they were sleeping. They woke up and said, 'There's water here, it's coming out at this point'. They threw away the straw and this icy water came out until daylight. That was the day they dug at the foundations of the hut. The water began to come out slowly, and as they dug, it increased. This is the water the elder asked for in his sanctification. They drew water at the foundations of the hut. This is why they are called the Xwâkwé (which means water source). They took this name because of this origin, this spring that gushed into the hut, that's it. When they came from Wânii, they created with another name, their ally was Néjia, it is said that Xwâkwé mê Néjia. Xwâkwé, who had stayed in that place, became the great Xwâkwé, that's what it is now".* (Noël Gouémoin, Xaracuu oral tradition 2003, p.s 50–51. Translation by the author of Xaracuu).

The ancient water, just like the wisdom of the elders, is primarily passed down through storytelling because it reveals traditions, such as the making of water containers, gourds and calabashes (cf. Sarasin, 2009), Kanak money, bows and arrows, water, and in general, all the individual traditions chosen for ancient water. This ancient water also refers to the Neolithic age of water and the movements that the young Kanak Bêrêda describes hereafter, because ancient water comes from the heights of the great rocks where it is stored. It is used not only by land travellers, but also by river or sea travellers, in pirogues, for them to consume after long journeys from one commune to another or from the Islands to Grande Terre.

7.4.2 The Young Bêrêda and the Ancient Water

> *Bêrêda handled and threw the ancient water, explaining the ancient water to us, the old water. You have to hold the ancient water and hit on the side with the other arm so that it launches into the air, flies before falling in front of someone like untouchable water then icy water. It's this water that makes the tradition happen, it's black, it rises and falls three times. You have to sit down and discuss the tradition of this ancient water, old water. The ancients called it that because it's old water from before the Neolithic period* (interview from 1979).

The elders also knew how to use cosmography to identify the Southern Cross, the Pole Star and the path of the winds. They waited for the Pole Star to descend so that they could collect it and store its water on the rocks. The appearance of the Pole Star is also important in children's water-sky circle games as well as for rituals that combine the magic of meteorological phenomena with cosmography. Ancient water is rare and depends on the stars and the sky, so it must be well managed and preserved. The ancient water was offered to a friendly gathering, and the sacred ritual delighted the crowds who came to admire it. In some places, the elder's property, where children showered with ancient water, was a space dedicated to cultural activities.

7.5 Collective Water

Collective water refers to the rivers or streams belonging to the mystical "masters of the land", whose clan stories are passed down to preserve their knowledge. These stories enable research to be carried out to improve the lives of the indigenous people. Other Kanaks, such as the spiritual leaders of thunder and rain, have understood it as mystical rather than collective water, and it is currently being studied with new considerations regarding the natural significance of water. Rules and limits are recommended, as there still remains much to be studied in terms of mythical events and figures. Kanaks often make the connection with cosmogony and cosmography before coming back to contemporary matters and water, which can help explain mythical events. Myths and legends therefore abound in stories about this water, to such an extent that the integration of mythical figures into the collective or mythical water must be debated, as they are not seen as masters of the land (for example, Noowa, the man of the legion or Wanewâà the fish woman) (Bensa and Rivière, 1994). Joachim Boanehotte, a small chief of the Weyem tribe, described this mystical figure as the owner of the Weyem River. These collective water stories are told on the main large and small rivers of New Caledonia (Weyem, Diahot, Hienghène, Tchamba, Négropo, etc.). However, the great stories told in films come mainly from the Hoot ma Whaap customary area, such as the legend of the Weyem River. Kaan Falik is a story book that recounts the appearance of mythical figures, including devils, in rivers in the Hienghène commune (Hoot ma Whaap) (Bureau du Patrimoine Culturel 2003).

The legend of the owner of the Weyem River (Joachim Boanehotte, small chief of the Weyem or Ouaième tribe, 2016):

> *I can't say it's a young man, an adult or a spirit, a deity or even a stone, in any case it comes from the source of the river. It may come from the bottom of the river, near 'Kuun nga' at a place called 'kuun we'. It's there where the stones are, for those who know the river well, it is said to come from there. Our ancestors used to say that the owner lived there. Nowadays, we think it's better to know our history well, so that people know the truth, especially the Kanaks and those who are coming. Let them know that we are indigenous. Because there are things that are forbidden that we know, but that we can't reveal; you can reach 'Kuun we' by boat and that's where the spirit comes from. That's why we say that the river is taboo. The elders used to say so. Sometimes its course is revealed by the appearance of bubbles and foam on the surface. You can tell where it's been by the bubbles! It's the path that the owner has traced to make its appearance. Sometimes it can be a fish, a ray or even a shark*

> *swimming upstream or downstream along the river. The elders used to say that when you see a fish, you mustn't stop it going up or down, or poke it – it's not a normal fish, it's the master of the river. These prohibitions all come from this legend. They say not to hunt because it's the spirit that has taken on an animal appearance. Sometimes it's a shark or a ray, it changes appearance. That's why the elders used to tell us not to catch rays and sharks. It's because of this legend.*
>
> *The legend of the man, the owner of this river. It is said that the river was made to be used, but if there is this taboo it is because of this spirit who lives there! Foreigners have heard this story. But we, the locals, know it, and we tell it to keep it alive. We all need to know. Everyone has their own story in their clan and where they live! We have to respect this word, because it's not our legend, it's the legend of the elders. It's to show that it's our story! They've told it to us so that we know it and so that those who come will know it too. That's how it's done... it's our strength, in our lives, in our clan, there where we stand. These words are our strength as natives* (Joachim Boanehotte, in Devillers, 2016).

For Martial Tyuénon, the spirits' path to these masters of the land is part of this tradition of natural and supernatural things, associating declamations with the things of time (Martial Tyuénon, Xârâcùù oral tradition, 2009, p. 42), just as Isabelle Leblic integrates the spirits and taboo places into the "Pays de surnature (Land of the supernatural)" (Leblic, 2005, p. 100).

7.6 Customary Land Water

The management of the drinking water supply in the Kanak tribes is currently under the responsibility of the water companies and the 33 municipalities in New Caledonia. This drinking water is referred to as customary land water, as it comes solely from the legal ownership of customary land guaranteed by the Constitution and it is not located on clan land. The vast majority of water catchments and storage in rural communes comes from customary land water: in Ponérihouen, for example, of the 38 catchment and storage facilities, 27 are located on customary land (Thétis Exploitation, 2015, p. 177). The term "water from customary lands" is mainly used in the documents of these water companies in order to attribute the ownership required to draw up work programs. The first drinking water supply systems were built in the 1950s through funding from the FIDES[4] (Kurtovitch, 2002, p. 215), the Indigenous Affairs Department, the Maréchaux des Logis (military), voluntary work by indigenous peoples and the public works department (Archives of New Caledonia, 97W/Trépied, 2011). Responsibility for works lies with the tribal chief and his clan council when the municipal waterworks are carried out on any of the clans' plots. This is guaranteed through written minutes of customary discussions taken by the gendarmes, as they know the history of the water used as drinking water in their tribes. Today, the government of New Caledonia is faced with the complex problem of how to invoice for water that is related to Kanak identity itself.

[4] Fonds d'Investissement pour le Développement Économique et Social (FIDES): Investment Fund for Economic and Social Development

> *The issue needs to be addressed with a certain degree of flexibility, so as not to offend sensitivities and to take into account the different ways in which space is perceived and experienced. For example, charging for water on the islands is perceived as an aberration. The same sentiment can be found on Grande Terre: why pay for a substance that is part of the Kanak's essence, of an identity that is existential to them, that is linked to their history and that of their clan?* (JONC, Customary Council, 2019).

For Kanak mayor Pierre Chanel Tutugoro, the current priorities are to improve water quality by renovating reservoirs in which the water is no longer drinkable, and to provide services to the tribes in his commune of Ponérihouen, which is still only partially connected to the network. This situation explains why families are still using the water supplied by the clans. His commitment to users led to the development of a master plan for drinking water supply in 2014–2015, backed by a budget dedicated to water issues. The fire brigade is also involved in securing water resources and by observing protective perimeters against pollution and deliberate bushfires: *"The installation of fire hydrants in the village for civil emergencies is one of the fire brigade's current priorities for dealing with fires. Additional actions, such as supplying drinking water to two families whose homes are disconnected from the public drinking water network, or education on civil safety, also show just how necessary water is"* (Pouyé David, fire chief in Ponérihouen, 2018).

7.7 Conclusion

Through this text, primitive forms of classification can be approached using collective memory, i.e. the five waters or the three lessons on the tradition of water. It also highlights the indigenous or mythical attributes of water through a wide range of activities and oral traditions (Muru-Lanning, 2018).

Clan water, legally recognised by the government of New Caledonia, shows the extent to which the Kanaks are attached to it, through the necessities of communal life and the beliefs associated with it. The water totem, the water gods and beliefs in the beneficial effects of rites involving astral or meteorological phenomena all contribute to reinforcing the status of water in the clans. It also highlights the spiritual leaders for water, thunder and rain, whose roles have yet to be fully explored by the social sciences. Although religiosity may have taken centre stage, it quickly becomes secondary to the fertility of the agricultural fields that provide for daily subsistence. The use of water, sun, rain and thunder stones has helped to forge clan and political relationships, by combating drought or, conversely, bringing back the sun, but they too have disappeared as a result of conflicts associated with the establishment of non-traditional religions.

Customary land water has been named as such by the water companies, in line with the constitutionally guaranteed ownership of customary lands. The incomplete nature of the texts has led to the use of clan water for general purposes as well as for civil security activities.

Pure water, flavoured water and water naturally sweetened with coconut are the drinks that Kanaks are renowned for. The water tradition allows for young Kanaks to be educated in how to make these drinks, in techniques and exercises for the body, as well as in irrigation techniques to cultivate taro.

Ancient water is located on the property of the elders and is located on tribal lands that are higher in elevation. Until recently, honouring the elders meant committing to an oral tradition or an individual tradition. It is considered to be the ancient water from the Neolithic period. Today, older people are changing the cultural activities on this property belonging to the elders, shifting toward trade or yam cultivation and to the detriment of preserving the ancient water, which is disappearing year by year.

Collective water is described as having a mystical owner. A study of the river is often carried out to inform the locals about the existence of mystical beings and the rules and taboos with regard to fishing.

These water classifications demonstrate the importance of the issues linked to each type of water. These issues and narratives reaffirm the status and roles that groups within the same political group have previously agreed to distribute among themselves and share.

This classification is also far removed from the classification traditionally used to manage the resource, and linked to hydrological functioning (for example, the distinction between surface water and groundwater). It therefore raises the question of how to account for this type of classification or these water concepts within a system for managing and preserving the resource, aimed at preserving water as a resource, but also water as a cultural or mystical entity.

Lastly, this classification sheds light on misunderstandings or tensions surrounding the payment for water or so-called "taboo" water (see Chap. 6): clan water is often claimed as "private" water located on clan land, collective water as "shared and common water", and ancient water as "taboo water". Finally, pure water is often associated with "good water", the water we drink, which is not necessarily tap water (see Chap. 8).

References

Aramiou, S., & Euritein, J. (2003). *Pèci i Bwêêyouu Ërijiyi*. Fédération de l'enseignement libre protestant (201 p.)

Bensa, A., & Riviere, J.-C. (1994). *Les filles du rocher Até. Contes et récits Paicî* (490 p.). Paris: Librairie orientaliste Geuthner-Nouméa, ADCK.

Bureau du Patrimoine Culturel (2003). *"Kaan Falik". Boutures de parole. Contes de la région de Hienghène*, Bureau du patrimoine culturel, n° 1, Province Nord.

Devillers, J.-L. (2016). Weyem, la vallée sacrée, Documentary film.

Garnung, F. (2016). *Contes et coutumes canaques au XIXe siècle* (80 p.). Paris: L'Harmattan, Collection la légende des mondes.

Gouemoin, N. (2003). le clan Xwakwé, tradition orale Xârâcùù.

Grochain, S. (2020). *Les cinq eaux kanakes* (155 p.). Report by the Institut Agronomique Calédonien.

Guiart, J. (1963). *La chefferie en Mélanésie du Sud* (688 p.). Paris: Institut d'ethnologie, Musée de l'Homme.
Herrenschmidt, J.-B., & Le Meur, P.-Y. (2016). *Politique foncière et dynamiques coutumières en Nouvelle-Calédonie et dans le Pacifique*. Province Nord, Océanide, IRD.
JONC (Journal Officiel de la Nouvelle Calédonie—Official Journal of New Caledonia) Customary Council. Notices of 19 and 26 February 2019.
Kona, R. (2007). Le mont Paabai, tradition orale Xârâcùù (Mount Paabai, Xârâcùù oral tradition).
Kurtovitch, I. (2002). *La vie politique en Nouvelle Calédonie: 1940–1953*. Lille, Presses universitaires Septentrion, thesis à la carte.
Lambert, P. (1999). *Mœurs et superstitions des néo-calédoniens* (367 p.). Nouméa, Société d'étude historique de Nouvelle Calédonie, No. 14.
Leblic, I. (2005). Pays, "surnature" et sites sacrés paicî à Ponérihouen, Nou-velle Calédonie. *Journal de la Société des Océanistes, 120–121*, 19. https://doi.org/10.4000/jso.410
Leenhardt, M. (1932). *Documents néo-calédoniens* (514 p.). Paris: Institut d'ethnologie.
Leenhardt, M. (1937). *Gens de la Grande Terre* (228 p.).Paris: Gallimard.
Leenhardt, M. (1947). *"Do Kamo" la personne et le mythe dans le monde mélanésien* (314 p.). Paris: Gallimard.
Leenhardt, M. (1980) (Reprint). *Notes d'ethnologie néo-calédonienne* (340 p.). Paris: Institut d'ethnologie, Musée de l'Homme.
Maramin, A. (2011). Les pierres, tradition orale Xârâcùù (The stones, Xârâcùù oral tradition).
Mauss, M. (1969). *Essai de sociologie* (246 p.). Paris: Édition de minuit.
Moasadi, P., & Diapoiri, E. (2008). Nényöwé, la tarodière irrigué, CCJMT/Customary council from the Xârâcùù area, 1 hr. 21min.
Muru-Lanning, M. (2018). At every bend a chief: Waikato of one hundred chiefs: Mapping the socio-political life of the Waikato River. In J. R. Wagner & J. K. Jacka (Eds.), *Island rivers: Fresh water and place in Oceania, Canberra, Australian National University*. (27 p. http://press-files.anu.edu.au/downloads/press/n4276/pdf/ch06.pdf)
Neba, J.-M. (2002). *Pierre de pluie, du tonnerre et du soleil, tradition orale Xârâcùù*. (Rain, thunder and sun stones, Xârâcùù oral tradition).
Ngoni, E. (2016). Les 10 ans des tarodières irriguées de la tribu de Mia, ADCK/CCJMT documentary film, 34 minutes.
Powi, J.-P. (2017). Récits sur l'eau, tradition orale *Xârâcùù* (Tales about water, Xârâcùù oral tradition).
Rognon, F. (2018). Maurice Leenhardt, Pour un "destin" en Nouvelle-Calédonie, Edition Olivetan.
Sarasin, F. (2009). *Ethnographie des kanaks de Nouvelle-Calédonie et des Îles Loyauté, 1911–1912*. Ibis Press. (297 p. 73–XIXpl).
Saussol, A. (1979). *L'héritage: Essai sur le problème foncier mélanésien en Nouvelle-Calédonie* (Vol. 40). Publication de la Société des Océanistes, Musée de l'Homme, , (493 p.). https://doi.org/10.4000/books.sdo.563
Thetis Exploitation (2015). Commune de Ponérihouen programme AEP 2015, Dossier de Marché, 200 p.
Tjibaou, E. (2007). Les tarodières irriguées de Uakatr, Video library at the Agence de Développement de la Culture Kanake/CCJMT, Customary council from the Iaai area.
Trepied, B. (2011). Des conduites d'eau pour les tribus. Action municipale, colonisation et citoyenneté en Nouvelle Calédonie. *Revue d'histoire moderne et contemporaine, 58*(4), 93–120. https://www.cairn.info/revue-d-histoire-moderne-et-contemporaine-2011-4. Documentary films and local documents
Tyuenon, M. (2009). Le chemin des esprits, tradition orale Xaracuu. (The path of the spirits, Xârâcùù oral tradition)

Chapter 8
Water Places: From Relational Ontologies to Water Governance

Catherine Sabinot, Pierre-Yves Le Meur, Mahée Autunno, Anne-Laure Dotte, Julien Drouin, Lucie Gosset, and Jacques Wamejonengo

Abstract "Water is the link" is an expression often heard during the interviews conducted in New Caledonia. Certain water places, such as springs, water holes and caves, are particularly meaningful for Kanak people and characterised by various uses. Understanding what they represent, how the inhabitants name them, describe them, and finally how they have been used and managed from the past to the present is a key step in thinking about water policies in New Caledonia. This chapter describes the diversity of values accorded to these water places, the assemblages of human and non-human actors they participate in, as well as the rules of use that apply to them. Focusing on recent surveys conducted among the inhabitants of Maré in the Loyalty Islands and Thio on the east coast of Grande Terre, we will analyse in particular the evolution of practices and norms with regard to freshwater supply sites, whether they are completely natural, partially or fully developed: water holes, caves, water springs and water cisterns. This chapter will show that a good understanding of these links is essential for designing a water policy acknowledging and integrating the diversity of ways of understanding and managing water in New Caledonia.

C. Sabinot (✉) · M. Autunno · L. Gosset
IRD (French National Research Institute for Development), UMR Espacedev (IRD, Univ Montpellier, Univ Guyane, Univ La Réunion, Univ Antilles, Univ Nouvelle Calédonie, Univ Perpignan), Centre IRD, BP 05, Nouméa, New Caledonia
e-mail: catherine.sabinot@ird.fr

P.-Y. Le Meur
IRD (French National Research Institute for Development), UMR SENS (IRD, CIRAD, UPVM), Centre IRD, BP 05, Nouméa, New Caledonia

A.-L. Dotte
UNC, University of New Caledonia, Equipe de Recherche Eralo, BP R4, Nouméa, New Caledonia

J. Drouin · J. Wamejonengo
IAC (New Caledonian Institute of Agronomy), Equipe SOLVEG (SOL et VEGétation), Station de Recherche Agronomique de Atha, BP 1407 La Roche, Mare, New Caledonia

C. Lejars, S. Bouard (eds.), *Water Policy in New Caledonia*, Global Issues in Water Policy 32, https://doi.org/10.1007/978-3-031-88196-1_8

Keywords Indigenous knowledge · Knowledge cocreation · Water governance · Nature connectedness · Relational ontologies

8.1 Introduction

"Water is the link" is an expression often heard during interviews in New Caledonia. Some water places such as springs, water holes[1] and caves, are places that have significant meaning and uses for the Kanak people. Understanding what they represent, how the inhabitants name them, describe them, and finally how they have been used and managed in the past through to today is a key step in thinking about water policies in New Caledonia.

This chapter, co-written by researchers in anthropology, linguistics and agronomy, having worked in different customary areas of Grande Terre and the Loyalty Islands, describes the range of values attributed to these water places, the interactions of human and non-human actors within them, and the regulations governing their usage. Focusing on recent surveys conducted among the inhabitants of Maré in the Loyalty Islands, Thio and Ponérihouen on the east coast of Grande Terre, we will analyse in particular the evolution of values, practices and standards with regard to fresh water supply locations, whether completely natural, partially or fully developed: water sources, "waterholes", sinkholes or underground caves and "cisterns". This chapter will explore how a good understanding of the links built around water is essential in order to devise a water policy that can recognise and integrate the diverse ways of conceptualising, managing and even caring for water in New Caledonia.

8.2 Understanding Water: Assemblages and Relational Ontology

Understanding water places requires a way of conceptualising water itself. The reduction of this element to a molecule, H_2O, makes it possible to identify other elements that disturb its composition, biochemical pollutants or metal contaminants. However, it does not allow us to grasp the place that water occupies in complex ensembles, composed of human and non-human actors, which give it meaning and purpose in a given socio-historical context (Linton, 2010). These ensembles can

[1] In a recent study (Richard et al., 2021), hydrogeologists and anthropologists have examined the lessons learned from both the physical and social sciences regarding the interrelationship between waterholes and the management of sedimentary heritage from mining in Thio, New Caledonia. Their objective was the enhancement of the inhabitants' involvement in the design and implementation of observations and monitoring of river systems.

be described as assemblages (Li, 2014), or as apparatuses (Agamben, 2009), if we want to emphasise the strategic and targeted dimension of their distributed action.

The preferred viewpoint for understanding water from a holistic point of view is that of a relational ontology, as it considers water's inherent human and non-human links and prioritises the relationships of interdependence and reciprocity (Ingold, 1996, 2018; Lescureux, 2006; Escobar, 2016; Sabinot and Lescureux, 2019) that make up water: a relational ontology "can be defined as one in which nothing preexists the relations that constitute it. Said otherwise, things and beings are their relations, they do not exist prior to them" (Escobar, 2016). In an unpublished text, Escobar added that in these ontologies, "life is inter-relation and inter-dependency through and through, always and from the beginning" (Escobar, 2013). Ingold (2018) states that in this sense "beings-in-relation are mutually constituted". The empirical material presented in this chapter emphasises these relationships and aims to qualify them more precisely by taking into account the diversity of their content and meaning.

In this context, water is conceptualised as an actor (or agent) among others, but occupies a central place. Water possesses agency, a term to be distinguished from intentionality: "If the actors within a given social system are human, non-human, animate, and inanimate, then agency does not have to be defined exclusively in terms of intentionality, but can be defined in terms of the impact of a given actor within the network" (Wagner, 2013: 7–8). The agency is also distributed among the different actors that make up a given assemblage (Bennett, 2010).

Water-centred assemblages define waterscapes, "the culturally meaningful, sensorially active places in which humans interact with water and with each other" (Orlove & Caton, 2010: 408). Swyngedouw (2015) speaks of *hydroscape* to underscore how water is inserted into complex material apparatuses while Strang (2006), with the notion of *fluidscape*, emphasises the specific materiality of water, its fluidity and its mobility.

The approach in terms of assemblages and water landscapes, expressing a relational ontology, allows us to think about the forms of hybridisation and tinkering in the sense of Lévi-Strauss in La pensée Sauvage (1962), taken up among others in recent works on water and its management (Benouniche et al., 2014; Peytavi et al., 2023) which describe "traditional" and "modern" water, in an obviously oversimplified way. This tinkering concerns both the daily practices of local actors and the development of water policies.[2]

Operationally, this approach is also reflected in the concept of a bio-cultural indicator (Gavin et al., 2015; Sterling et al., 2017), the potential of which we will explore in the last part of this chapter. Additionally, it opens up a more "horizontal" and "symmetrical" vision of the relationship with water, which covers, for example, the notion of water care as opposed to the more top-down approach based on the notion of control, of water management (Domínguez-Guzmán et al., 2022).

---

[2] See also Chap. 13.

8.3 Water Places, Places of Value Markers of Continuity Among Entities

What specifically is a water place? The question is simple but the answer can be complex if we try to comprehensively define it by considering all its different dimensions.

In Kanak languages, there are many terms for naming water, real water, river, water hole… In the Paicî language, the rivers naming method is in itself particularly illustrative of the fundamental values that are given to water places. In this Kanak language of the Grande Terre, the river is called *nâiriwâ*. *Nâ* means "extent of", "set of", "in"; *iri* means "skeleton" while *wâ* means "house". The river is understood as the skeleton, the structure of a shared living space, inhabited by men and women as well as by non-human animal, plant or mineral entities or artefacts. In the Kanak view of the world, water is considered to be the only "natural path" that unites beings, living and dead.

In each of the 28 Kanak languages, there is a term for talking about water[3]: in the Loyalty Islands *tini* in Nengone, *wai* in Fagauvea (or its free variants *tai~vai*), *köiö* in Iaai, *tim* in Drehu; in Grande Terre *jawé* in Paicî, *kwé* in Xârâcùù, *we* in Caac and the languages of Hienghene, *tabè* in Cèmuhî, *wii* in Nêlêmwa, etc. However, one does not always find in these languages a specific term to designate rivers or streams, in coherence with the unique term for 'fresh water' and 'river', 'spring' which is reconstructed in Proto-Oceanic (the common ancestor of all the Kanak languages): *waiR (Ross et al., 2003: 58). The Kanak languages of Grande Terre mostly reflect this, and confirm that fresh water may have been, in the first place, more thought of as drinking water that emerges rather than flowing water; "no Proto-Oceanic term has been reconstructed for 'river' as distinct from the term for 'fresh water'. Speakers of Oceanic languages would probably lack the map-based view of a river thought of primarily in terms of an entity with length. Rather, they seem to conceptualise it simply as fresh water that flows." (ibid. 60). The diversification of languages over time, and in particular Kanak languages, has changed this semantic division of origin by bringing new lexical forms to new ecological realities and new representations of environments.

In addition, there are many qualifiers to describe water in Kanak languages: It is sweet or soft (*tini buruiaa* in Nengone, *köihnam* in Iaai, *kwé niöru* in Xârâcùù, *wii nam* in Nêlêmwa, *nreconrii* in Naa Drubea) to designate spring water, bitter/salty(*tini me kece* in Nengone, *köihiny* in Iaai, *kwé mâdù* in Xârâcùù, *wii yaak* in Nêlêmwa) for seawater, brackish/cloudy for water at the river mouth (*kwé poro* in Xârâgurè). Water from waterholes formed by karst collapses in the Loyalty Islands is also lexically differentiated (*(tini)wapep* in Nengone, *pupua* in Fagauvea and Iaai for water that springs from the earth). Water is also qualified by its temperature (hot

[3] All these terms are listed in the Xârâcùù and Paicî languages in the reports of Gosset (2020) and Autunno (2022), otherwise in the existing dictionaries of Kanak languages.

or cold),[4] its colour (*tai epu* 'turbid water' in Fagauvea, white when it is moving, red when it is loaded with red soil, like *vaimelo* in Fagauvea), its quality (good or bad, like *tai maalie* 'good water' in Fagauvea), its potability as drinking water or not (it is often referred to as "real water", as *duu we* in Pije and Fwâi, *ju-tabè* in Cèmuhî) and its spiritual or "sacred" character (*pwiciri* in Paicî). A water hole can thus be described as *nä-jawé pwicîrî* in Paicî, literally sacred water hole marked by restrictions (Leblic, 2005). For those who have been converted to Christianity, holy water is called *kwê xiri* ou *kwê xiti* en Xârâcùù. However, the qualifier *xiri* is first commonly used to speak of a place on which it is necessary to respect rules (prescriptions and restrictions) of access and use that vary according to the place and the person, a place then often referred to in French as "taboo" or "sacred". In the next section, we will describe how such a place is organised and how its management has evolved over the past decades.

Prior to that, we will focus on describing how water places are inhabited, the relationships that shape them and the links that bring them to life. Indeed, there are many "water places" that have a strong symbolic and customary value, linking the invisible world of the dead with that of the living, the different entities among themselves and the humans among themselves.

Discussions around water places quickly showed that for the tribes' inhabitants from Thio, Ponérihouen or Maré, water's meaning extends beyond the concept of a resource. Many forms of continuities exist between water and environmental elements (plants, animals, minerals, humans). The meaning given to water by Kanak society must therefore be understood "beyond water", taking into account the links of continuity that exist between this resource and any other element, and are part of a relational ontology.

In Kanak way of conceiving the world, everything works in paired associations of complementary symbolic elements (Leblic, 2005; Sabinot & Herrenschmidt, 2019; Tjibaou, 2020; 1976). For example, women and water, taro and woman, fire and water, dry and wet, stone and water, eel and mullet are linked. These are only a few of the associations including the element of water, which is always at the heart of the links, as being the origin of life. Similarly, clan genealogical discourses work in pairs (Pillon, 1992; Naepels, 1998).

During our investigations on the link to water and water places in Grande Terre and the Loyalty Islands, four forms of relationships resonating with the definition of "beings-in-relations" that are "mutually constituted" (Ingold, 2018) were identified: association, circulation, kinship, and correspondence/analogy. These links express the way Oceanian peoples conceptualise and order the world, which does not involve a nature/culture dichotomy, nor land/sea discontinuity (Sabinot & Herrenschmidt, 2019). In this way, water acquires meaning when it is part of assemblages formed by these four main types of links.

---

[4] In the northern languages of Grande Terre, a different verb is used depending on whether you drink a hot liquid or a cold liquid (Dotte & Moyse-Faurie, 2021: 214).

8.3.1 Associations of Entities Inhabiting Water Places

In the Loyalty Islands and on Grande Terre, several animal, plant and mineral species are associated with water places. They have a role in maintaining water in the water place, both in terms of quantity and quality.

Eels are present at river headwaters in Grande Terre and are often associated with black mullet that live in rivers and like to settle in water holes and springs; eels and shrimps inhabit large water bodies, whether they are deeper water sections of the rivers of Grande Terre or caves or karst sinkholes filled with water in the Loyalty Islands, such as on Maré. These species are considered the "guardians" of the place, to keep the water at the source and to ensure the abundance of the resource in the water holes and in the creeks. Catching them in too large quantities may lead to the water drying up.

In the St-Paul mountain tribe in the Thio commune, the expression *xa pwara kwé* is used in Xârâcùù to describe the harmony existing between the water and its environment involving the presence of the eel and meaning that "everything is in its place". In this same tribe as in some tribes of Ponérihouen, it is a specific species of shrimp, a "tiny shrimp", which plays a key role in maintaining the water place: *pâè kakö* in Xârâcùù, its scientific name is *Macrobrachium caledonicum*:

> *All the old people always said, that they [the tiny shrimps] along with the eels and all, that we don't touch them, because the eel is just to maintain the flow, the flow of water. It is the eel who maintains the water table and all that. Pasè kakö, its role is to clean the inside. (Man from St Paul tribe, Thio, 2020).*

Other testimonies from Ponérihouen relate that the absence of certain species explains why the water place is poorly maintained:

> *The water is dirty, especially with algae, it's as if the fish, black mullet, eels and shrimp could no longer clean up their habitat."; "It's like a house. [...] when the person leaves, we know that the house is abandoned, because there are spider webs and everything. Here, it's the same.* (Man from Amoa tribe, Poindimié, 2022)

The plant elements associated with water are also signs of harmony and some are used by humans to manage the flow of water. For example, niaouli bark can be used to filter water from a source, plug a dirt pipe, restrict a flow and thereby reinforce a catchment. Some tree roots such as those of the ironwood (*âbwia*, as this tree is called in Xârâcùù) will not always be removed during maintenance of a pipe in order to limit the passage of leaves and dirt.

Mineral elements play a vital role in water places and in the interactions they allow with the invisible world and other human and non-human entities.

On the one hand, they serve as landmarks, as place markers: The "blue pebbles" of the waterhole of the twins or the *Nèdè xiri* in Saint Pierre or the ones marking the location of the place of the spring in Saint Paul (Thio). When a water hole is particularly important and meaningful, one of its slopes and a part of the bed are generally made of a single rock, considered by locals as the "mother earth" connected to the ancestors (Pêchîato water hole in Thio). This rock is then responsible for the formation of whirlpools and eddies (*ngêdùrùu* in Xârâcùù) which enable the water holes to maintain a specific depth and serve as habitats for species of high cultural significance, such as black mullet or eels.

Finally, some stones, commonly known as "pebbles", are used by knowledgeable people who are able to influence climatic phenomena or behaviour of fish. In water places there is a "rain pebble" which can bring rain or a "black mullet pebble" which can increase the population of mullet, as reported by the man interviewed in Saint-Paul (Thio) in 2020: "*there was a pebble for this fish, to multiply it you know. It is a pebble but it takes the form of that fish. For there to be a lot of fish, they will put it in a small creek, they will secure it well, and the water will pass over it, and according to them it will then it will multiply [the number of fish].*"

8.3.2 Circulation is Central to the Existence of Water Places

Only certain people from families or clans have the power and position that enables them to activate the stones to guide the species of fish in their movements or in their reproduction or those to make the water spring up, to channel water into a place by using a piece of wood that is considered sacred. A link of continuity between water and humans is therefore clearly established, some even testifying that when water clans move, "the spring follows; it follows the person"[5] or that the former dwelling places of the clan elders retain a certain humidity. In Cèmuhî, the name *aapwihî a tabè* means 'boss, water magician'.

The topic of circulation came up in all the interviews conducted about water places. It is both the circulation of water itself, transported or guided by humans and invisible spirits, as well as that of riparian species between different locations.

The previous story shows that water can circulate guided by a man and his "medicine" infused stick[6]: "*Because the water passes under it, the water follows it. If he [the old man] stops, the water comes out where he stops*" says a man from St Paul. Water can therefore circulate from one water place to another simply by following the course of a river or by taking underground passages invisible to the living beings.

Many people also report that wildlife circulates according to its reproductive cycle, with some species moving up and down the river to reproduce and feed. Some fish are also known for their ability to pass from one spring or waterhole to another via underground channels located at the bottom of certain waterholes such as in Ouvéa or Grande Terre, also called "mullet holes" by the locals. In the Saint Yves tribe (Ponerihouen), the only way to dive into the deep and dirty water hole, starting from the underpass to La Tchamba, is to go there with "medicines". The movement of riparian species between two water places marks the link between the territories

[5] Comment from a Saint Philippo 1 (Thio) man, 2020.

[6] Term commonly used to speak of a visible and invisible set of element that carry magical, therapeutic powers, sometimes contained in a plant "package".

that house these water holes. In particular, they reveal the relationships linking human groups to each other.

8.3.3 Water Places Marked by Relationships of Ancestry, Parentage and Identification

The ties of association and circulation complement each other with another form of relations marked by relationships of kinship, whether parentage or alliance, and which are sometimes interpreted in terms of identity relationships. As mentioned above, the link to water and water places is based on people belonging to a specific clan whose function is passed down from generation to generation. These people can move water; they are also responsible, as we will see, for access to and maintenance of water places. The stories, which were shared with us about how a source of water had emerged in this or that place, were based on the presence of families linked together by blood ties or by bonds of marriage. This is how moving water to a place allows water access to be maintained for descendants and also results from the act of gifting in the context of marriages.

Some unions and lineages are even metaphorically associated with the image of the water source invigorating relations between clans, as seen in Cèmuhî with the expression "*tabè duu popwöli cuö hé bèt, tabè bwala hê ni mwa me ni amu* 'spring gushing into the crowd, fresh water of clans and countries' (to designate a peacemaking lineage in ceremonial style)" (Rivierre, 1994: 339).

We can metaphorically extend the notion of kinship to that of guardianship. There are springs, water holes, and caves for which there is a gender and age-specific responsibility. The Saint Paul spring in Thio is thus called *Pêê mè paxûchî,* literally "the woman and the children" marking a link of fertility and parentage with the place and implying dependent rules of access. In addition, both shrimp and black mullet species known to be linked to living water and quality water places are also considered to be "mother of all shrimp" and "mother of all fish" (freshwater and seawater), respectively. It is also known that the black mullet named *dénâ* in Paicî, present in Ponérihouen, spawns in the sea and gives birth to marine species in a kind of totemic logic specific to a group of non-humans.

The term "totem" is commonly used in French in New Caledonia but in a rather inconsistent manner. This is partly due to the fact that investigations are mostly conducted in French. The black mullet is considered as the totem of certain clans but sometimes, the magic stone associated with it, is itself called a totem, in a way that can be described as metaphorical. In a simplified manner, totemism can be understood as a classification system (Lévi-Strauss, 1972), a religious cosmology (Durkheim), or an ontology (Descola) based on the idea of a physical and spiritual connection between groups of humans and non-humans. Without going into the details of this complex debate, we can be guided by Aletta Biersack who states: "Totemism rests upon two affiliations: first, between the human species and other species and, second, between religion (particularly ritual) and society" (Biersack, 1999: 70). The emphasis is on a relationship of identification, if not identity, between humans and non-humans. This is also what Sahlins reminds us when he describes totemism as segmentary animism: "Totemism is segmentary animism, in the sense that different non-human persons, as species-beings, are substantially identified with different human collectives, such as lineages and clans" (Sahlins, 2014: 282). In 2023, customary representatives from the customary senate working to define and recognise a Kanak cultural vision of the ocean opted for the term ancestor species, more faithfully reflecting the fundamental relationship of the Kanak way of "being in the world" (Sabinot & Herrenschmidt, 2019).

The links between non-humans as well as between humans (clans) and non-humans (animals, plants, but also natural entities such as thunder) also relate to strong relationships of ancestry, parentage and identification. The fact that this worldview endures in narratives, even in Ponérihouen where religion has significantly altered perceptions of the world, indicates that Kanak society has retained a significant part of its pre-colonial intangible heritage. It underscores the importance of integrating these connections with the world into a collaborative water policy to ensure the continued preservation of water places.

8.3.4 Water Connections of Correspondence and Analogy

In the Kanak way of conceiving the world, connections between human and non-human entities are viewed through associated pairings. Some of these connections are based on a symbolic correspondence link that exists between several entities. Water and the places where it emerges (such as springs) or places where it has a fertility role with a strong identity value (such as taro fields) hold a fundamental place in this correspondence. Water, women, springs and taro fields are all synonymous with fertility and the source of life.

These forms of correspondence relationships generate a number of taboos regarding women's access to and contact with water, such as at Saint Paul mother spring, especially during certain periods of their life associated with the menstrual cycle. The taboo relates to women's ability to cause a drying up of the water place and influence fertility. So, the woman finds herself balancing her creative and destructive abilities which can both impact water places.

While women's contact with water can have effects on water, it can also impact women and influence their own fertility. It is known by residents that a pregnant woman bathing in certain water places such as the "water hole of the twins" in Thio is more likely to give birth to twins. Conversely, bathing in other water places such as the *Né pwé mwaa* waterhole (literally "the belly of the house") can lead to sterility.

Finally, the stem of the water taro, symbol of water, life and women, is used in the preparation of purges for newborns in Kanak communities (especially in Thio,

Touho and Pouebo) before they drink breast milk, imparting new breath, new life to the child.[7]

8.4 Practices Governed by Management Standards That Consider Relationships Between Clans, Between Humans and Non-humans

The multiplicity of values accorded to water places has implications in terms of practices, rules of access and standards for the maintenance of each place. Acknowledging the diversity of ways of thinking and acting around these places aims to feed into reflections on the institutionalisation of water management and therefore to support the development of a water policy capable of integrating the diversity of perspectives and place-specific management.

In this section, we will revisit the various forms of continuities that give rise to distinct assemblages, which must be acknowledged to effectively manage water places and resources.

Next, we will outline current practices observed in these regions, covering both water usage and associated resources, along with the management procedures relevant to these locations.

8.4.1 Assemblages Resulting from Different Forms of Relations

The different assemblages linking water to other human and non-human entities combine to produce arrangements that convey values, knowledge, know-how and standards. Figure 8.1[8] shows the complexity of these assemblages and makes it possible to understand the extent of the social and cultural value represented by water places.

The responses collected across different sites in New Caledonia are unanimous: "Water is the link".

We must consider that it is the links between the different human and non-human entities that define water. In designing a shared water policy, it is therefore important to consider the assemblages formed by these links in caring for water places.

[7] For a more detailed approach to water flow in an individual's (intimate) life cycle, see Chap. 9.

[8] Figures 8.1 and 8.2 have been modified from Gosset et al. (2020).

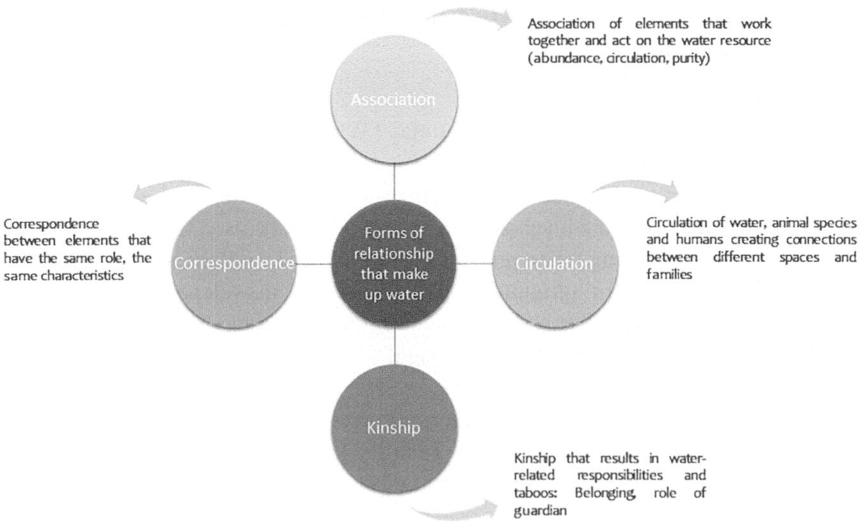

Fig. 8.1 A schematic representation of the forms of relationships linking water and water places to other living entities, both human and non-human

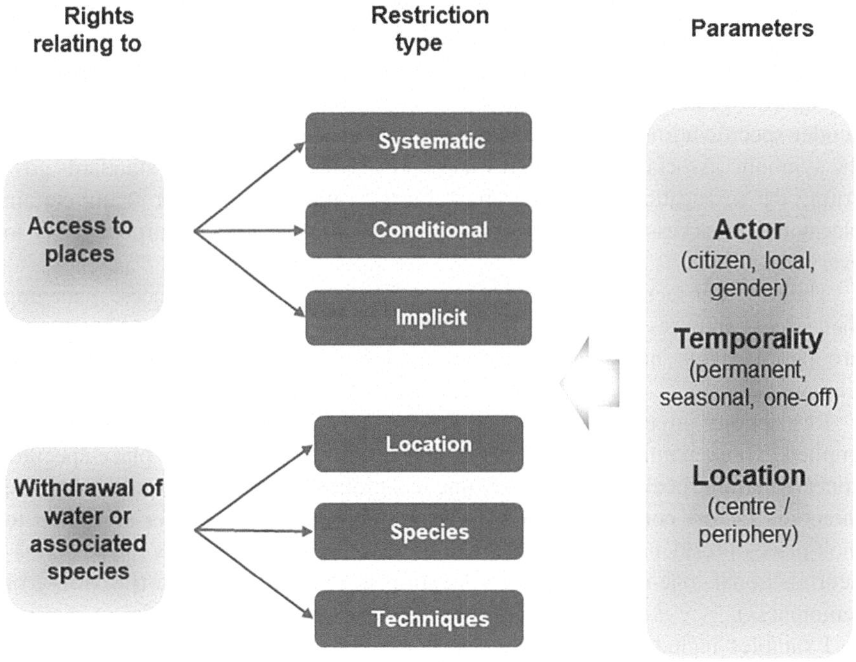

Fig. 8.2 Key factors taken into account for caring for water places in the Kanak environments

8.4.2 Standards of Access and Use Embedded in a Relational Ontology

How are water places used and managed today? What practices and standards of use result from these assemblages?

A water place can, depending on its geographical location, and as long as the local standards associated with each place are respected, be used for drinking, treating illness, fishing, rinsing when returning from the sea, cooling, washing laundry, or even drawing water for agriculture. The diversity of uses depends on the place and time of the year, as well as the socio-symbolic dimensions specific to each place and the forms of relationships that comprise it.

During our investigations, we focused on describing and analysing the practices and standards of access and management associated with water places such as water holes, springs and caves. This led us to identify key criteria specific to each site: temporality of practices, status and number of people that may be affected, associated access and management rules.

While it is common for no specific rules to exist for the water holes "on the periphery" of the tribes and where there is no attachment, those close to the tribe and dwellings are managed by at least two levels of rules: rules of access and drawing water for persons outside the tribe, and rules for the inhabitants of the tribe (closure of fishing in water holes at certain times, for example). This is the case in Saint Paul for the two water holes located opposite the tribe (*Bwaa mâsî* et *Pê chîa*) and in the tribe of La Tchamba where two water holes have been banned from fishing for two years to allow for black mullet reproduction. Furthermore, there are gender-specific norms that result in restrictions on access for women or men due to the symbolic associations between water and women, or due to the standards governing the segregation of bathing areas. Lastly, rules as to whether or not certain places can be accessed at night or the resource used for a given purpose apply to everyone.

The different factors taken into account for caring for water places, managing their access and uses are summarised in the diagram above, represented in terms of groups of rights and access restrictions following (Ribot & Peluso, 2003) frameworks.

Restrictions governing access to the site may be systematic, conditional or implied. Those regulating uses (water or fish withdrawal) may be place-specific, concern certain species or some sampling techniques. For each of these restrictions, three key factors come into play: The actors concerned (who differ according to their place of residence, gender or function); temporality (restrictions can be permanent, seasonal, one-off); the place's location in relation to a tribe (proximity or remoteness).

Examples include permanent access restrictions for everyone to a spring and certain water holes (*Né pwé mwaa*,[9] for example), restrictions on women's access to

[9] Water hole located upstream of the village of Thio, in a Caldoche hamlet nicknamed "Dallas".

springs, waterholes or taro fields subject to their menstrual cycle or if they are pregnant, seasonal restrictions on fishing for all following a decision of a person or group having authority over the place. Access to certain bathing or fishing places located within a tribe or linked to a particular clan may be contingent on performing a customary gesture or a specific request and sometimes involves inhabitants from outside the tribe being accompanied by people from that place (this is the case, for example, for access to the *Raçöö* and *Bwaa mâsî* water holes of the Saint Paul tribe or for access to Pombei Falls, formerly open to people outside the tribe). Finally, there are implicit restrictions that fall under "good behaviour" rules. The most illustrative example is that of not fishing in a water hole where another fisher is already actively fishing.

The rules of access and withdrawal are not exclusive and can coexist in the same water place. The reason for the restrictions may be symbolic or territorial in relation to persons outside the tribe. The goal is to preserve water places and water and fish resources, and more generally to maintain relations between humans and non-humans, and to take care of the world.

An interview excerpt from the Ponérihouen commune provides a particularly good illustration of these interactions between humans and non-humans, including elements such as the sun, thunderstorms, pebbles, fish and humans. The person spoke in 2022 about a taboo water place called *Tiboboa*, a water hole located under a small waterfall: "*You take the rock, you throw into the water but you must not stay there, because the sky will darken, the thunder will rumble and burst, the flood comes [...] You can bathe [in the water hole], it is not forbidden. You can catch fish and shrimp there. But do not shake the water. Basically that's it. But there is a place higher up called Nétcheboé. This place is also a taboo place, the water hole of thunder, there when someone agitates the water, you see the sky close in, you must immediately break a branch and then it will be ok, the sun, the light will return. Until now, it's always been like that.* The man from Ponérihouen who tells this story emphasises the strong interactions between human actions and the reactions of a place, from heaven to earth, often affecting the people responsible for the place or the action performed.

8.5 Uses and Standards Possibly Arising from the Combination of Traditional and Modern Practices?

How can this diversity of values, uses and standards be integrated into the rules for the protection and maintenance of water sites? How does a water policy consider water at the country, province or municipality level in a way that is defined in a relational way in the context of assemblages linking it to other human and non-human elements?

The knowledge and know-how that human has developed in their relations with water, both in the past and present, are diverse and complementary. Bridges can be established by integrating Kanak actors and their roles into the teams responsible for maintaining rivers and water networks. This integration is occasionally already happening, as communal hydrant men often belong to water clan (Peytavi et al., 2023) or considered "hybrid water masters" where this clan role has been lost, such as on the Yandé Islet in Poum (Budna Perrin, 2021: 45). The standards of use described above are actually standards of responsibility for living water. The various forms of continuity involve forms of responsibility in which men, women, water clans, and all local inhabitants have a role to play.

8.5.1 Living and Inhabited Water Related Responsibilities

The responsibilities for maintaining a water place flow from the living nature of the water, from the place and what inhabits it, from all the connections that make up water. The rules of access, restrictions and taboos are all an expression of a necessary form of respect.

The terms most often heard about a place to be respected are the words 'taboo' and 'guardians', who can be human as well as non-human (Godin, 2020). Ancestors and invisible beings have an essential role in these places and living beings are in regular contact with them to ensure proper maintenance of water places (Sabinot et al., 2020; Tjibaou, 2020).

The practices described in the previous sections are contemporary, not forgotten values or norms. Despite criticisms and occasional breaches of certain aspects of these management standards by certain individuals or groups, they remain largely prevalent overall. Sometimes, however, it is the materiality of assemblages centred on "traditional" water that has been altered by mining pollution, land dispossession, deaths of individuals responsible for the place or displacement of populations, leading to loss that affects both the places and the associated knowledge and norms.

Certain events such as heavy cyclonic rains may have led to massive and sometimes sudden pollution of watercourses, profoundly transforming the practices and management methods associated with certain sites (Gosset et al., 2019). For example, Tomuru Creek in the Saint Paul tribe was completely transformed (engravement, hypersedimentation, widening by several metres) whereas up until the 60s, it was the location of an open channel built by the inhabitants of the tribe from the spring to the taro fields and dwellings. This led to the gradual cessation of the cultivation of water taro, the installation of a "modern" pipeline to supply the tribe with water, and the end of collective management methods for the open channels. These events also strongly question the quality of drinking water distributed by municipalities.

In Maré, we asked some interviewees to classify the water, from the best to the worst. The most common answer was: "The water from the water holes [caves] is the best"; then that collected from the trunk of the coconut trees and the brackish water from the seaside. The latter are water places named *ura* in Nengone, fresh water resurgence flowing to the surface of the beach. Their location is transmitted from generation to generation and between young people who work and play in the fields and at sea. It is always good to know where it is possible to find some water to drink when returning from fishing, hunting or working in the fields. In Ouvéa, rainwater stored by each family is preferred to the water supplied by the commune which comes from the seawater desalination plant (and that must be paid for). On the east coast of Grande Terre, the water distributed by municipalities sometimes contains heavy sediment after the rains and consumers periodically prefer to use spring water collected by hand or rainwater. The preferred water is that which has circulated naturally, which is guarded and carried by living beings, whether fish, shrimp, pebbles or human.

Several recent scientific publications discuss the role of guardians of water places, sections of rivers, springs and water holes on Grande Terre (Lejars et al., 2019; Coulange et al., 2021). In Maré the most noticeable and remarkable water places are the karst collapses resulting from limestone erosion that has formed extensive underground water networks and chasms, gradually infiltrated by rainwater. They are commonly known as sinkholes, sometimes located in caves. While their significant role in providing drinking water nearly vanished with the arrival of missionaries, establishment of wells, and subsequent development of water supply systems, the residents remain familiar with this practice. Access is often restricted to a few families or clans who act as "guardians". This is the case for the *Trou de Bon* in Maré, belonging customarily to the Guama chiefdom, which is guarded by the Wakuela family (Sabinot et al., 2020). Access is free today, but water can only be collected by one of the family members. Previously, the water was taken via a *gubon*, a braided basket made of banana leaves weighed down with a stone.

The former guardian we met described the responsibilities associated with this place and spoke in particular in 2020 about the transmission of this responsibility, when he passed on his "customary power" to his son three years earlier:

> *I handed over power to him because he is already married. When the eldest is married we automatically pass on the power at all levels, the yam festival, the land, everything... he got married in 2016, I passed the power to him in 2017, we always leave a year after the marriage. His role in relation to the water holes is the guardian. We are the guardian of course, but there is a great chief who is above all else. The role of the water place guardian is to ensure that no one misbehaves or throws anything into the water, and supervise tourists when they visit. They have to come here to be given authorisation. They can be chased out of there, if they don't have permission. For a few years now, the great chief has asked us to allow access to promote tourism.*

This lengthy excerpt also highlights the role of chiefdoms in the Kanak social structure, an organisation where everyone has their specific role. However, the evolution of practices and norms is always possible to facilitate tourism development, as seen here, or to assume the responsibilities typically assigned to an absent individual or a defunct clan (even if these arrangements have their limits and some clan roles may no longer be performed). In any case, the question of responsibilities associated with water management, how to take care of it and respect the links and assemblages that sustain it, is not solely an individual concern, it is anchored in the hierarchies and roles that shape Kanak society.

8.5.2 Hybridisations and Tinkering in daily Water Management

In the daily management of water, we have always observed hybridisations of knowledge or tinkering as defined by Lévi-Strauss (1962). This occurs by drawing on diverse types of knowledge, that of the elders, that acquired by regular experience of the environment or that proposed by other communities, institutions or the younger generation. These tinkerings also know how to "work" with the constraints arising from contexts and events that have introduced disruptions – such as land expropriations, mining expansions, the establishment of infrastructure, or new policies—within the pre-existing flow of environmental management and knowledge.

In the Loyalty Islands, which has no rivers, many creations have emerged to collect or conserve water. In Maré, on coconut tree trunks, you can still see *wabanu*, hollows made "in the belly of the coconut tree" to collect the water that flows down the trunk guided by coconut leaves knotted around it. Each *wabanu* can hold 3–4 L of water that was in the past mainly used as drinking water. They are found near houses but also scattered in fields far from houses. Cisterns formerly called *wie* can also be found close to houses. They are produced by the transformation of limestone from the coral reefs that form the island's substrate.

Given the lack of surface water on the Loyalty Islands, these creations to harvest small or even large volumes of rainwater are very precious and once again became the topic of conversations during the long El Niño season accompanied by significant droughts and very regular water cuts that can last several days or even weeks.

A man testifies about the use of *wabanu* and also *wie*: "*In the past, every place where there was water was extremely precious. So we also took care of these places. Each time I went with my mother to remove the leaves, the roots. The rocks were always very clean.*"

Maintaining water places, whether they are under the responsibility of a few or several clans, or now managed by a more institutionalised figure such as a communal hydrant man, remains a major challenge. It is essential to care for these places to sustain all forms of relationships with water, ensuring the water remains alive. We see a form of institutional tinkering (Cleaver, 2012) on Taanlo, an islet in the Poum commune. The person in charge of maintaining the desalination unit established to supply the locality with water plays the role of "master of hybrid water", being both Kanak from this place and communal agent (Budna Perrin, 2021: 44–45).

Regarding the precious nature of water and the significant value attributed to this resource in the past, several people interviewed emphasised the importance of educating youth about traditional practices: "*Young people won't necessarily know the importance of water if we don't tell them how difficult it was for us to get even a few litres of water.*[10] That water is the link and made up of links means it must be respected and preserved throughout all its interactions. This is an important message to share with Kanak youth as well as all the inhabitants of New Caledonia, especially since the future will probably bring more intense periods of drought than those the country has previously experienced (Bailly et al., 2023).

For a commune or chiefdom, ensuring that its population has access to drinking water can be achieved by developing and maintaining the water supply network, taking into account the various uses of the water and the population's perceptions of what constitutes good drinking water, whether it comes from a spring, a river or rain (see Chap. 9). It can also be based on the development or even the reactivation of old water retention practices by recreating dams along a river, by promoting agricultural practices that retain water, or, as in Maré, by restoring some rainwater cisterns such as *wie*, or even constructing new *wie*, which a little chief of a Maré tribe did in 2020, as evidenced by the conversation excerpt below:

> *A man from Maré: "In each tribe, there was at least one cistern. Now we no longer see any cisterns, only a few remain. Even if they are still there, they are no longer used [...].*
>
> *Investigator "The little chief asked to recommence construction of the cisterns?"*
>
> *Maré man: "Yes, and the province is also interested in replacing some cisterns (renovation, construction)."*

In March 2024, a Loyalty Islands Province official explained that he had indeed taken up this issue. He reports that renovations are underway on several of the Province's islands and that tests have been undertaken to ensure quality treatment at a reduced cost for these individual water reserves. In this situation, we can see that various levels of territorial governance intersect to secure access to fresh water in a municipality lacking rivers. In addition, during the 2020 interview, the man from Maré also indicated that while rainwater storage was no longer practised since the water network had been installed, several people, while worrying about the reactivation of abandoned dug and limed reservoirs, had decided to purchase plastic cisterns (300,000 L). This "new trend", to use the words of our interviewee, was put in place following the severe drought spanning several years that affected all of New Caledonia until 2020 due to the unusual duration of El Niño, leading municipalities to regularly cut off access to the public water network.

---

[10] The full excerpt concludes with why it is important not to link charging for water with accountability. The complex issue of water payment, which cannot be simplified to the commodification of a natural element, is dealt with in Chap. 13.

8.5.3 *Taking Care of Water Today: Human Relationships and Monitoring*

Taking care of the water and today's water places can easily be done by relying on traditional practices and promoting modern tinkering through diverse tools and methods, reflecting varied perspectives on water. This can also be achieved by bringing on board Kanak actors who already have a role in the maintenance and circulation of these water places, as we have just specified.

Exploring or reconstructing the diverse forms of relationships that define the essence of water remains crucial at all times to envision suitable and proactive management strategies. Different forms of knowledge exist, and while they may not be accessible to everyone, it is not possible to consider water management without accommodating this diversity or leaving room for the inventive tinkering that works around various constraints (Peytavi et al., 2023). Bringing diversity together at the tribal, municipal, provincial or country levels will obviously be one of the challenges of a shared water policy.

Finally, to monitor the water's movement, the quality of the resource and the links that sustain it, it is not enough to rely on biological or technical indicators capable of providing information on the biological or biogeochemical load or turbidity of water. It requires everyone to be included in monitoring efforts and possibly engaging the population in constructing bio-cultural indicators alongside other more conventional indicators for monitoring water quality. Among these indicators, "cultural keystone species" are identified by Garibaldi and Turner (2004) as crucial elements in fostering monitoring and conservation ambitions. Bio-cultural indicators are at the interface between ecosystem health indicators and socio-cultural indicators on norms and values. The former aim to measure different biochemical, physical or sometimes ecological parameters. They are not always understood by residents and can be very expensive to acquire due to the instrumentation required. The latter are often built to assess networks and regimes of norms and knowledge, or even the well-being of individuals and communities, and rarely include environmental variables. The notion of bio-cultural indicator lies at this interface. In addition to allowing co-opted monitoring and understanding by a diversity of actors, bio-cultural indicators are designed to facilitate dialogue between local populations, scientists and technical services. They are built by bringing local values, knowledge and know-how alongside scientific knowledge. Surveys conducted in 2022 and 2023 in the communes of Thio, Ponérihouen and Hienghène (Autunno et al., 2023) showed, for example, that the inhabitants of the mountain tribes have excellent knowledge of how black mullet populations evolve in time and space, because of the high value placed on these species. They observed population fluctuations and their movements and identified different environmental and anthropogenic causes for population declines. They believe that black mullets are sensitive to biophysical changes in the river (rising temperatures, pollution, hyperdimentation etc.) and recognise the role black mullets play in the health of the river. A team worked to identify how fish that are highly meaningful in Kanak environments, such as mullets or eels, are markers of a form of relationships by association and parentage, can be bio-cultural indicators to monitor river water quality and the quality of human-non-human relationships.

If water management experts and stakeholders embrace this proposal[11], by acknowledging and integrating the diversity of knowledge and knowledge producers, as well as the broader social, cultural, and environmental context of managing human and natural systems (Sterling et al., 2017; Dacks et al., 2019), then the shared water policy in New Caledonia will be enriched by considering water more in line with local conceptions and ways of caring for and monitoring the quality of this assemblage involving humans and non-humans. This will also mean not seeking direct equivalence between types of knowledge (for example by reducing indigenous knowledge to "naturalistic knowledge") because these are embedded socially, culturally, normatively in different ways depending on whether they are scientific, indigenous, professional or legal for example, but rather recognising their differences and complementarities.

8.6 Conclusion

Water has an intrinsic value, or more precisely it fulfils a vital function by meeting the physiological need of our bodies. It also holds significant relational value, tied to both human and non-human actors and events, embedded within assemblages with varying compositions and diverse temporalities.

Therefore, when establishing a management approach for a water place, irrespective of spatial or social scales, it can only be understood, recognised, and implemented if these varied values and relationship dynamics are considered during negotiation and subsequently integrated into management protocols.

This recognition presupposes an openness to the multiple arrangements and tinkerings that make up the ever-evolving assemblages that transform around water. These reflect various forms of articulation, juxtaposition or blending of different conceptions, norms, technologies and infrastructures available. In addition to highlighting the richness of relationships surrounding water and water places, involving humans as well as non-humans, the diversity perspectives and actions around these water holes, caves and cisterns, we aim to have offered insight to all stakeholders involved in designing water policy that will facilitate the integration of diverse approaches to understanding and managing water in New Caledonia.

[11] The proposal was discussed at a workshop involving various stakeholders in water management in New Caledonia in 2023, and was the subject of a specific sheet in the Freshwater Guide produced the same year (Sabinot et al., 2023).

References

Agamben, G. (2009). *What is an apparatus?* Stanford University Press.

Autunno, M., Bouard, S., Coulange, D., Le Meur, P.-Y., Richard, D., Despinoy, M., & Sabinot, C. (2023). *Pour une gestion intégrée et inclusive du milieu ripicole en Nouvelle-Calédonie : une étude des connaissances, des représentations, des valeurs et des usages associés aux mulets noirs sur la côte Est de la Grande Terre* (87 p.). (Rapport final du projet Mulet noir financé par le CNRT "Nickel & Son Environnement).

Bailly, F., David, V., Ganachaud, A., Houlbreque, F., Lehodey, S., Le Meur, P.-Y., Menkes, M., Sabinot, C., Soler, N., & Vallet, F. (2023). La recherche scientifique face aux défis du changement climatique. *Revue juridique, économique et politique de la Nouvelle-Calédonie, 41*, 145–156.

Bennett, J. (2010). *Vibrant matter. A political ecology of things*. Duke University Press.

Benouniche, M., Zwarteveen, M. Z., & Kuper, M. (2014). Bricolage as innovation: Opening the black box of drip irrigation systems. *Irrigation and Drainage, 63*(5), 651–658. https://doi.org/10.1002/ird.1854

Biersack, A. (1999). The Mount Kare Python and his gold: Totemism and ecology in the Papua New Guinea Highlands. *American Anthropologist, 101*(1), 68–87.

Budna Perrin, M.-E. (2021). *L'eau plurielle, une résistante à l'enclavement dans une commune minière (Poum, Nouvelle-Calédonie)*. Master thesis in anthropology. EHESS.

Cleaver, F. (2012). *Development through Bricolage. Rethinking institutions for natural resource management*. Routledge.

Coulange, D., Bouard, S., Grochain, S., Lejars, C., Sabinot, C. (2021). Analyser et comprendre les représentations, les savoirs et les pratiques autour de l'eau sur terres coutumières pour appuyer la mise en oeuvre de la Politique de l'Eau Partagée (PEP). In : Perrier-Gras A. (ed.). L'eau dans tous ses états en Nouvelle-Calédonie. Revue Juridique, Politique et Économique de Nouvelle-Calédonie, 37, 207–222. https://hal.science/hal04418476v1

Dacks, R., Ticktin, T., Mawyer, A., Caillon, S., Claudet, J., Fabre, P., Jupiter, S. D., McCarter, J., Mejia, M., Pascua, P., Sterling, E., & Wongbusarakum, S. (2019). Developing biocultural indicators for resource management. Conservation Science and Practice, 1(6), Article 6. https://doi.org/10.1111/csp2.38

Domínguez-Guzmán, C., Verzijl, A., Zwarteveen, M., Mol, A. (2022). Caring for water in Northern Peru: On fragile infrastructures and the diverse work involved in irrigation, Environment and Planning E: Nature and Space 5 (4): 2153–2171, https://doi.org/10.1177/25148486211052216

Dotte, A.-L., & Moyse-Faurie, C. (2021). Toward a comparative typology of "Eating" in Kanak languages. *Oceanic Linguistics, 60*(1), 199–230. https://doi.org/10.1353/ol.2021.0006

Escobar, A (2013). Notes on the ontology of design. Unpublished Paper. http://sawyerseminar.ucdavis.edu/files/2012/12/ESCOBAR_Notes-on-the-Ontology-of-Design-Parts-I-II-_-III.pdf

Escobar, A. (2016). Thinking-feeling with the Earth: Territorial Struggles and the ontological dimension of the epistemologies of the south. *AIBR, 11*, 11–32. https://doi.org/10.11156/aibr.110102e

Garibaldi, A., & Turner, N. (2004). Cultural keystone species: Implications for ecological conservation and restoration. *Ecology and Society, 9*(3), art1. https://doi.org/10.5751/ES-00669-090301

Gavin, M. C., McCarter, J., Mead, A., Berkes, F., Stepp, J. R., Peterson, D., & Tang, R. (2015). Defining biocultural approaches to conservation. *Trends in Ecology & Evolution, 30*(3), 140–145. https://doi.org/10.1016/j.tree.2014.12.005

Godin, P. (2020). Interdits et tabous. Lagarde, L. (ed.). Le patrimoine de la Nouvelle-Calédonie, Editions Hervé Chopin, Fondation Clément, p 492–493, 2020, 978-2-3572-0534-37. https://hal.science/hal-04922489

Gosset, L. (2020). *Les lieux d'eau remarquables de la tribu de Saint-Paul, Thio*. Research Report, Nouméa, IRD, Goutte Project.

Gosset, L., Sabinot, C., & Worliczek, E. (2019). Quand cyclones, pluies et pollution interrogent les liens des Kanak à leurs rivières et participent au renouvellement des savoirs écologiques (Thio,

Nouvelle-Calédonie). In *Ethnographiques.org, Numéro spécial Approche anthropologique des changements climatiques et météorologiques*. https://doi.org/10.25667/ethnographiques/2019-38/004

Ingold, T. (1996). Hunting and gathering as ways of perceiving the environment. In *Redefining nature: Ecology, culture and domestication* (pp. 117–154). Berg.

Ingold, T. (2018). *Anthropology. Why it matters*. Polity Press.

Leblic, I. (2005). Pays, «surnature» et sites «sacrés» paicî à Ponérihouen (Nouvelle-Calédonie). *Journal de la Société des Océanistes, 120–121*, 95–111.

Lejars, C., Bouard, S., Sabinot, C., & Nékiriaï, C. (2019). Quand «l'eau, c'est le lien»: Suivre l'évolution des réseaux d'eau pour éclairer les pratiques et les transformations sociales dans les tribus kanak. *Développement Durable et Territoires, 10*(3). https://doi.org/10.4000/developpementdurable.15704

Lescureux, N. (2006). Towards the necessity of a new interactive approach integrating ethnology, ecology and ethology in the study of the relationship between Kyrgyz stockbreeders and wolves. *Social Science Information, 45*, 463–478. https://doi.org/10.1177/0539018406066536

Lévi-Strauss, C. (1972). *The savage mind (La Pensée sauvage)*. Weidenfeld and Nicolson.

Li, T. M. (2014). What is land? Assembling a resource for global investment. *Transactions of the Institute of British Geographers, 39*, 589–602.

Linton, J. (2010). *What is water? The history of a modern abstraction*. University of British Columbia Press.

Naepels, M. (1998). Histoires de terres kanakes. Conflits fonciers et rapports sociaux dans la région de Houaïlou (Nouvelle-Calédonie). Paris, Belin, 380 p., bibl., lexique, tabl., cartes.

Orlove, B., & Caton, S. C. (2010). Water sustainability: Anthropological approaches and prospects. *Annual Review of Anthropology, 39*, 401–415.

Peytavi, O., Bouard, S., Le Meur, P.-Y., & Lejars, C. (2023). Freshwater supply as sociotechnical tinkering: The co-creation of water knowledge and assemblages in New Caledonia. *Journal of Political Ecology, 30*(1). https://doi.org/10.2458/jpe.5289

Pillon, P. (1992). Listes déclamatoires (viva) et principes d'organisation sociale dans la vallée de la Kouaoua (Nouvelle-Calédonie), Journal de la Société des Océanistes, 94(1), 81–101.

Ribot, J. C., & Peluso, N. L. (2003). A theory of access. *Rural Sociology, 68*(2), 153–181.

Richard, D., Sabinot, C., Gosset, L., Worliczeck, E., Pidjo, A., Bertrand, M., & Garcin, M. (2021). Towards a better integration of local populations in the design of projects to manage the massive aggradation of rivers stemming from mining activity in Thio, New Caledonia. *The Extractive Industries and Society, 8*(1), 74–73. https://doi.org/10.1016/j.exis.2020.08.011

Rivierre, J.-C. (1994). *Dictionnaire cèmuhî-français, suivi d'un lexique français-cèmuhî*. SELAF Peeters.

Ross, M., Pawley, A., & Osmond, M. (2003). *The lexicon of Proto Oceanic. The culture and environment of ancestral Oceanic society/Volume 2: The physical environment*. Pacific Linguistics.

Sabinot, C., & Herrenschmidt, J.-B. (2019). La dynamique des pratiques change-t-elle la manière de penser les relations kanak aux continuités terre-mer et à la nature? *Anthropologie et Sociétés, Alternatives locales à la conservation environnementale*.

Sabinot, C., & Lescureux, N. (2019). The local ecological knowledge and the viability of the relations with the environment. In O. Barrière, M. Behnassi, G. David, V. Douzal, M. Fargette, T. Libourel, M. Loireau, L. Pascal, C. Prost, V. Ravena-Cañete, F. Seyler, & S. Morand (Eds.), *Coviability of social and ecological systems: Reconnecting mankind to the biosphere in an era of global change* (The Foundations of a New Paradigm) (Vol. 1, pp. 211–222). Springer International Publishing. https://doi.org/10.1007/978-3-319-78497-7

Sabinot C., Drouin J., Wamejonengo, J (2020). Les cénotes. Lagarde, L. (ed.). Le patrimoine de la Nouvelle-Calédonie, Editions Hervé Chopin, Fondation Clément, p. 516–520, 2020, 978-2-3572-0534-37. https://hal.science/hal-04922489

Sabinot C., Autunno M., Bouard S., Coulange D., Despinoy M., Le Meur PY, Richard D., (2023) Fiche 12 : Les indicateurs bioculturel : les mulets noirs, des poissons suivis par les riverains des cours d'eau. In Dominique Y., Mardhel V., Mackenzie S., Winckel A., Gunkel-Grillon P., Boissard G., Russ L., Cereghino R., Delmas F., Roset N., Touron-Poncet H., Côte C., 2023.

«GUIDE DES EAUX DOUCES, Guide de suivi de la qualité environnementale des eaux douces en Nouvelle-Calédonie », Rapport Final. 218 p.

Sahlins, M. (2014). On the ontological scheme of beyond nature and culture. *Hau: Journal of Ethnographic Theory, 4*(1), 281–290.

Sterling, E. J., Filardi, C., Toomey, A., Sigouin, A., Betley, E., Gazit, N., Newell, J., Albert, S., Alvira, D., Bergamini, N., Blair, M., Boseto, D., Burrows, K., Bynum, N., Caillon, S., Caselle, J. E., Claudet, J., Cullman, G., Dacks, R., et al. (2017). Biocultural approaches to well-being and sustainability indicators across scales. *Nature Ecology and Evolution, 1*(12), 1798–1806. https://doi.org/10.1038/s41559-017-0349-6

Strang, V. (2006). Introduction. Fluidscapes: Water, identity and the senses. *Worldviews, 10*(2), 147–154.

Swyngedouw, E. (2015). *Liquid power: Contested hydro-modernities in twentieth-century Spain*. MIT Press.

Tjibaou, J.-M. (1976). Recherche d'identité mélanésienne et société traditionnelle. Journal de la Société des Océanistes, 32(53), 281–292. https://doi.org/10.3406/jso.1976.2754

Tjibaou, E. (2020). Les mondes souterrains. Lagarde, L. (ed.). Le patrimoine de la Nouvelle-Calédonie, Editions Hervé Chopin, Fondation Clément, p. 510–515, 2020, 978-2-3572-0534-37. https://hal.science/hal-04922489

Wagner, J. R. (Ed.). (2013). *The social life of water*. Berghahn Books.

Chapter 9
Water Quality on Customary Land: An Intimate Link Between Humans and Their Environment

Olga Peytavi and Emmanuel Tjibaou

Abstract This chapter examines the concept of "good water" within Kanak society, exploring the intricate relationships between humans and their environment to highlight an alternative water cycle that diverges from institutional perspectives. It integrates discussions with Emmanuel Tjibaou and customary authorities from the Cèmuhî and Xârâcùù regions. This chapter highlights the necessity of redefining water management strategies to include indigenous knowledge systems. The concept of "good water" in Kanak society provides a holistic understanding that encompasses the physical, social, and spiritual dimensions of water. Water is viewed as a continuous flow through the body, landscape, and society, sustaining life and maintaining harmony. Conversely, murky water signifies disruption and disharmony. By embracing this comprehensive view, water management in Kanaky/New Caledonia can be significantly improved, leading to more resilient and adaptive practices that benefit both the environment and the community.

Keywords Kanak · Water · Indigenous knowledge · Ethnography · Intimate link

> There is life, quality, things everywhere, even in a water hole at the bottom there are eels. When there is no quality, it means that it is polluted, that there is no more life. The question of quality doesn't make sense because it's a question of physical composition. You have to understand the water and what's around it. "Everything is connected". (Interview with Xârâcùù customaries, Water Forum, Poya, April 2022)

This chapter traces the trajectories of the relationship between humans and their environment to demonstrate an alternative water cycle that diverges from institutional perspectives, and the specificities of what is "good water" for Kanak people. In this chapter, the focus on water relations goes beyond material and physical

---

O. Peytavi (✉)
IAC, CIRAD, IRD, Centre de Recherche Nord-Thierry Mennesson, Pouembout, New Caledonia
e-mail: olga.peytavi@cirad.fr

E. Tjibaou
Culture Department, Province Nord, Kone, New Caledonia
e-mail: e.tjibaou@province-nord.nc

C. Lejars, S. Bouard (eds.), *Water Policy in New Caledonia*, Global Issues in Water Policy 32, https://doi.org/10.1007/978-3-031-88196-1_9

relations, but is embedded in the social and historical context of alliance networks. By exploring the conceptions of water quality and their implications for drinking water distribution methods and the organisation of sanitation, we will show in this chapter that water quality is also a way of reading human relations, which goes well beyond the water we drink! Water is attached to the place we live and its quality cannot be detached from it. This quality is also linked to the history of water and of relations between humans and non-humans, where relations of power or dependence are played out. Quality makes it possible to retrace the path of these relationships and bring to light the transformations and ruptures that underlie the frictions around water in Kanaky/New Caledonia.[1]

Before moving on to the notion of quality on customary land, we will look at studies of water quality or its perception in the Pacific. In the literature on the subject of freshwater quality in the Pacific,[2] there are articles from socio-hydrology and management sciences that describe the quality of the water system in growing Pacific cities. These papers report on social inequalities within cities and between urban and rural areas (Anthonj et al., 2020; Moglia et al., 2008; White & Falkland, 2015), on environmental or human impacts on river quality (Carpenter & Lawedrau, 2002; Mosley et al., 2004), reflections on water quality management (Roumasset et al., 2010), as well as several reports on water quality monitoring in the different Pacific islands (Hughey et al., 2010; SOPAC & GWP Consultants, 2007; World Health Organization & Regional Office for the Western Pacific, 2005). These reports had a section on the perception of water quality by the inhabitants. We therefore focused our search on articles that dealt with the perception of water quality by Pacific residents, but these are restricted to empirical questions of taste, colour or odour, which complement techno-scientific knowledge of molecules or indicators measured (Crampton & Ragusa, 2016; Hughey et al., 2010; Nelson et al., 2022; Robak, 2022). Focusing on Kanaky/New Caledonia, a number of studies have been carried out to understand the water supply and sanitation system. In Kanaky/New Caledonia, several water-related research studies have been conducted since the start of the PEP in 2019 (DAVAR, 2019), including the GOUTTE project carried out between 2019 and 2020. One of the objectives of this project was to understand the representations associated with water on customary lands, through the links to man,

[1] The term "Kanaky" emerged during the protest movements of the 1970s, in opposition to the term "Nouvelle-Calédonie" imposed by the French colonial state since colonization. Today, while this term has not been officially recognized, it is predominantly used by the indigenous population in everyday language to refer to their country. The authors have therefore chosen to use the dual term "Kanaky/Nouvelle-Calédonie" in reference to this vernacular usage. This is especially relevant as the use of vernacular names is common in Pacific islands, such as the term "Aotearoa" used by researchers in Maori studies to refer to New Zealand.

[2] The Pacific Community (SPC) is the principal scientific and technical organisation in the Pacific region, supporting development since 1947. Its mission is to help Pacific Communities in developing sustainable well-being according to Pacific culture, environment and resources. They develop numerous projects, including the PROTEGE (Projet Régional Océanien des Territoires pour la Gestion durable des Ecosystèmes) program in Kanaky/New Caledonia, around freshwater use to unsure a sustainable governance of this natural resource.

the land, language-culture and the sacred (Bouard et al., 2020; Grellier, 2019). This project highlighted the singular relationship of Kanak populations to water (Coulange, 2020; Grellier et al., 2019; Lejars et al., 2019), as a link to the land, a link between families, clans and tribes, a link to stories and myths (Coulange, 2020; Coulange et al., 2021; Gosset, n.d.; Gosset et al., 2019; Grochain, 2020; Levacher et al., 2017; Nekiriai, n.d.). This chapter further explores these connections between humans and water. It focuses on the notion of quality as an intimate link between humans and water.

9.1 What Is Water Quality? Ontological Differences

In 2020, the Directorate of Health and Social Affairs (DASS) conducted the Health Water Barometer study across Kanaky/New Caledonia, financed entirely by the SPC (see footnote 2). This study is based on just over 1500 people. In order to have a representative sample of Kanaky/New Caledonia, a minimum of 30 people were interviewed per municipality (see Chap. 4). This is the first study on the scale of the island, an initial assessment of water perceptions among inhabitants: How does the population perceive tap water, rainwater, and how do they envisage the health risk linked to the water they consume, and finally how do they see themselves in the future maintaining this resource? One of the objectives of this study is to prompt inhabitants to change certain practices in relation to their use of water, knowing that the DASS insists on the fact that the water resource is in danger as we evolve over time. DASS sees this study as a snapshot in time, to facilitate an initial identification of needs, in order to work on awareness-raising and more targeted communication. DASS would like to hire a psychosocial psychologist to create appropriate communication tools with the population. However, progress on this has currently stopped due to lack of funding.

The detailed DASS study shows that the inhabitants of Touho rate the water at 12/20, which is below the general average of 14/20 for the whole of Kanaky/New Caledonia. Indeed, when the inhabitants and the hydrant men were questioned about the water in the network, it was often described as cloudy. While the majority of homes are connected to the network in the Vieux Touho, Touho-Mission, Kokingone, Pouiou, Tiwae, Kongouma and Amoa (we mention these tribes because they are the ones we visited and where these comments were reported to us), the water from the tap often comes out cloudy, as if there were a slight film of soil or dust. When opening the tap, water first flows out milky with little bubbles. The inhabitants told us that this is due to the pressure and indicate that they let the water run until it becomes clear. However, they rarely drink the tap water directly, but boil or freeze it first. They alternate between using tap water for the flowers in front of the house and rainwater to irrigate the fields of yams, cassava, kanak cabbage and other tubers and leaves.

In their study, the DASS uses terms such as: drinking water, treated water and health risk. These words are part of the DASS lexicon and refer to a very specific perception of water, one that is mainly shared by the institutions, and in which water is seen to be measured, standardised, regulated and above all detached from its environment and reduced to its molecular composition, H_2O. When we use these terms in the field, we quickly run out of points of discussion. So, what terms are used by the Touho inhabitants and how do they talk about water and its quality? When we talk about water, very quickly the old people will start telling us stories. What if we took it seriously? What if instead of starting from what water quality is, and try to apply it in the field, we turn the question around and start with vernacular terms and in particular the term 'good water' or 'puela jawé'[3] to understand what this term means for Kanak people.

Two contrasting views of the water cycle emerge: the drinking water cycle and the water cycle of life portraying the links between clans and between the elements of the environment. We will emphasise the ruptures and transformations that lead to this conception of water. Access to drinking water is a universal right, and is one of the major issues at stake in public policy. This article therefore aims to enrich reflections on the way in which we inhabit places and therefore access water, in order to promote inclusive intersectoral policies that integrate the diversity of populations on the territory by reducing inequalities.

9.2 Water, Life and Flow

9.2.1 Water Is Life: A Social and Spiritual Link Between Humans and Non-humans

During my research, customary people kept repeating the sentence "Water is life", the constant repetition of this phrase made us look into it more seriously. Water is life, a rather common statement that is not only specific to Kanak society, it underlines the importance of water as a resource that enables life, linked to the first microorganisms, the first cells that populated the earth. In this way, water is described as an indispensable resource, but as an element separate from humans, which they use to live. However, this first explanation does not satisfy us, as the use of the statement by the customary people seems to have a different meaning. The first time we understood its importance was at the water forum in Noumea in March 2022. One of the members of the Sénat Coutumier closed the morning session with "Water is life". He told us to address the problems faced by those with low incomes, in the tribes in remote areas, as well as those living in squats in the outskirts of the city. He ended his speech by saying that the debates on water must be directed towards these social inequalities, because water is life. I encountered this sentence again a few days later during the custom at the Water Forum in Poya: The mayor of Poya began his speech by asking the chief of the customary area to accept his presence and to allow the

[3] Puela jawé in paicî means clear water and can be used as an expression meaning "good day?"

Water Forum to be held here on his territory. The chief of the area thanked him and accepted his custom, stressing that water is an issue of the utmost importance because "water is life". These multiple recurrences lead us to take a more serious look at the phrase, "water is life", what do these different people mean about water, about life. What connection do they make between these two terms? Is it only an additive link in that you need water to live? Or is it a more intimate link in which water and life are inseparable. Water is no longer a resource or an element separate from humans, but an intimate gateway into life, revealing the connections between the human and non-human world.

9.2.2 *Water Flows and Individual Intimacies*

Inner intimacy: A continuous flow through the body. "But the way of us, men and women, is, that we are going to dry up."

Just now, I was joking with Pierre-Yves, I said to him: "But you've grown old! You're drying up." Whereas you two are alive, where there is water. But the way of us, men and women, is, that we are going to dry up. (Interview with Emmanuel Tjibaou, April 2022)

More than a simple metaphor, in retracing the words of my interviewees, water retraces the life cycle of an individual, from birth to death. As a physical component of the body, my interviewees define it as diminishing over the course of one's life, leaving marks and traces on the body, just as human life flows through wrinkles, hair and teeth, all of which help to build and dry out the body. Speaking of birth and the infant, Emmanuel Tjibaou takes the example of the infant's food, initially liquid, which solidifies little by little as it grows. Emmanuel Tjibaou draws a parallel between the infant filled with water:

And when you arrive, at birth, well, you have no teeth. You eat porridge, you eat water. And I give this as an example in relation to the idea of birth, of life, profusion of water. And then, the older you get, the drier you get. (Interview with Emmanuel Tjibaou, April 2022)

In his speech, Emmanuel Tjibaou traces the path of water through the physicality of people, flowing at the same pace as life. We note the first hair, menstruation as physical signs that allow people to detach themselves from the water they carry and to grow physically as well as socially during these moments of initiation.

And in fact, you go up, you reach the stage of initiation only because you let go of that part, which is the water. You dry up, you harden. (Interview with Emmanuel Tjibaou, April 2022)

In this last extract, detaching oneself from water allows one to become a human in society. While the loss of water dries humans up and inevitably leads them to death, it also allows them to harden. Water is therefore linked to the physical intimacy of the person, a vehicle of life within the body.

Outer intimacy: A continuous flow through the world. "So it's those who are on land and those who are below"

And the putrefaction flows into the creeks, which brings it to the mouth of the river. And at the mouth of the river, it starts the cycle again, transformed into clouds that go back and forth in the water. And so on. (Interview with Kokingone customaries, Touho, July 2022)

Water not only connects the human to its inner physical intimacy as the water that flows through the body, but also to its outer physical intimacy. The human life cycle is also connected to the cycle of water flowing from the mountains to the sea and back to the mountains through rain. This intimacy between the physical body of an individual and the external environment helps to erase the bodily limits between human and non-human. The body envelope and that of the environment are superimposed and water acts as a vital fluid that flows through its multiple envelopes and participates in their transformations.

The sea, the salt water, represents the realm of the dead into which the souls come to throw themselves after having lived on earth. Emmanuel Tjibaou establishes a correspondence between salt water and the tears one cries at a funeral:

Regarding quality: You see for example when you are mourning. The tears that flow or "wé le thamaan",[4] it's the water that flows. This is the custom you do when you arrive at someone's house, and the dying person is there. You go to pay your respects. Well, that's what we do, it's because it's salty (referring to tears). But it's already the water from the other side, the sea that you're crying after him, who has gone down the path, towards the underwater depths. So it's also a quality that is attached to one of the properties of water: clean water, salt water, pure water as I said earlier. (Interview with Emmanuel Tjibaou, Tjibaou Centre, April 2022)

Here, the salt water associated with the realms of the dead is manifested in the tears, drops of salt water that run down the cheeks of the deceased's relatives. These tears reveal the physical continuity between human and non-human and help to blur the bodily boundaries.

- But between freshwater and saltwater, there is. It's a, it's a continuity, or ? [...] well, it continues, but at the same time it changes its nature a little because it becomes salty?
- Yes, and it becomes a vital fluid that is attached to the underwater spirits, whereas fresh water is attached to us, the living. So it's those who are on the land and those who are below. That's why it reminds me of that too. The history of the cycle of life.

[4] Literally water/of/eyes.

- It's interesting because we could approach the quality of water, as if water is good, not good, and so on. But in fact, water is also an intrinsic part of the body.
- Well, we're on top of it. We are part of the world in fact. We are part of the universe, we are not dissociated from the thing. Maybe we'll say "OK, I can appreciate the quality of the water but I'm part of it so I'm responsible". (Interview with Emmanuel Tjibaou and Pierre-Yves Le Meur, Tjibaou Centre, April 2022)

In this discussion Emmanuel Tjibaou is defining salt water as a type of water quality, but which likewise cannot be separated from freshwater, it changes its nature but remains the same vital fluid. He links it directly to the cycle of water flowing from the mountains to the sea, then evaporating into the air, to the life cycle of a person going to death to return to the spirit realms. He stresses the role of water as a link from the human world to the spirit world. To support Pierre-Yves' reflection, which rephrases Emmanuel Tjibaou's idea that water is an intrinsic part of the body, the body is also an intrinsic part of water. The human body is intimately connected to water. The different qualities or properties of water allow us to define the different intimate links that humans have with their environment. Salt water here corresponds to loss, death and to the spirit world.

The notion of responsibility that Emmanuel Tjibaou evokes at the end of this extract underlines this intimacy and holds humans responsible for the world they are a part of and cannot separate themselves from. We find a movement of reciprocity, of give and take, where the human makes up the world, just as the world makes up the human.

In this section we have therefore been able to trace the intimate physical link between humans and their environment. Quality, while being an indicator of good water, also highlights the nature and composition of the link between human and non-human. Through these extracts we also begin to see that it is not only a physical link, but that there are also social links that we will describe in the next sections.

9.3 Water, a Social Organisation and Transmission

9.3.1 Water During Life Ceremonies: Transmission Through Birth, Marriage and Death

Talking about water, well that's life. When we look closely at our genealogy, we always talk about the environment, we talk about water. Why do we talk about water? Even when there is a customary death, we always talk about water. Because, for example, I am here today, but the transmission was made by water. From father to mother, it's water, it's life. And when you go to the sea, it's salty, but it's still life: woman and man, that's water. We are talking about the land and sea environment. That's why the water from the mountain always

> feeds downwards, yes. We Kanak rule like that, that's the Kanak world. (Interview with Kokingone customaries, Touho, July 2022)

In this extract we see the connection water establishes between human life and the water cycle from the mountains to the sea. Added to this intimate link are the key notions of genealogy, custom and transmission. The life of an individual is punctuated by customary ceremonies, also referred to as life cycle ceremonies (Monnerie, 2012). The three main customs being: birth, marriage and death. They are part of regularly events of a person's life, starting with their own, but also those of their more or less distant relatives. These customary ceremonies are part of the Kanak custom.[5] Custom is a living system of rules and kinship that manifests in everyday life and is used by the Kanak population to position themselves in the human and non-human world. Having a place in the custom means that each person has rights and duties towards other members of society according to the kinship ties they have with them. For example, the inhabitants of the Touho commune address each other on a daily basis according to their kinship ties. The kinship system is based on a "irokois" system, where the naming of individuals related to Ego differs depending on whether they are related through a man or a woman[6] (Godelier, 2010). Through these customary gestures and words, the inhabitants of Touho constantly enacte their place and their link in society and the material environment they live in.

During the three customary events, the elders recite the genealogy orally. Through the genealogy they retrace the path of the clans, starting from the mountain, the river, the stream they came from to be standing here today. One of the customary officials of Touho evokes the term retransmission, because these customary events open the space for the elders to recite the genealogy and for the children to listen to it and learn it. They are times and places to learn about their history and to be able to recite it in the future, when it is their turn. The wedding is a good example of the links between two families, two water streams woven together.

> We were dancing pilou to celebrate the union of the two clans. The two clans that were separate, by the man and by the woman, are now united. So a new creeper, a new manou, a new watercourse, it's a new stream that's linked. (Interview with Kokingone customaries, Touho, July 2022)

If water is important symbolically because it allows the history of the clans to be traced, it is also important materially. Indeed, these events are surrounded by water: a wedding or death custom is spread out over about a month to welcome relatives who have come from afar. During this time, the house of the parents of the bride and groom must be able to accommodate them, lodge them and prepare food. In

---

[5] It is challenging to have a single and clear definition of what Kanak custom is because it is solely oral and alive. It manifests itself in diverse ways depending on the time and place. However, we can emphasise that it is a system of values and rules based on exchanges and gifts of objects, words, and gestures (Bouard & Sourisseau, 2010). It enables the connection of social and environmental relationships (Bretteville, 2018).

[6] The mother's sisters are called "maman" (mom), and the father's brothers are called "papa" (dad), while the mother's brothers are referred to as "oncles" or "tontons" (uncles), and the father's sisters are addressed as "tantes" or "tantines" (aunts).

practice, there has to be running water to accommodate everyone. Water, like other daily duties, is managed collectively. Each family brings their own cooking pots which they share during the month of marriage celebrations and has access to the territory of the family hosting them to help with fishing, hunting and gardening. Water is made available to all. In Touho municipality, we were told of several situations in which water ran out due to water cuts during a wedding. The families then went to the river or the sea to wash the cutlery, and on one occasion in Tiouandé, north of Touho, the town hall sent the fire brigade with jerry cans of water, as the water ran out for several days. Water is therefore not only linked to the intimate life of an individual, but as can be seen from the customary ceremonies, it is also linked to the collective life of individuals. We propose to examine this intimate collective relationship, rooted in genealogy.

9.3.2 *Water and the Organization of Society: Naming One's Place*

9.3.2.1 Settling Along the River

> Yes, you have to talk about water, depending on the movement of each clan. Because they will go and sit down, they always go and look for water where there is water. Afterwards, that's where the branches leave from. The other one leaves here, the other one leaves on that side and the other one continues on the path. This genealogy, well, it's going to start from the water points. That's why in genealogy, if there is water, there is life. It's life, that's how I see the path of life, the life of all the clans. (Interview with Kokingone customaries, Touho, July 2022)

Through the words of my interviewees, we understand that water cannot be separated from the physical links between humans and their environment, nor from the social links that water weaves between humans. In this section we will focus on further exploring this intimate social link. If water represents life, the history of a human, it also represents the life of a clan, its migratory journey from its original mound to the place where they are settled today. This intimate link to the environment, which bears the traces of the passage of clans, has already been mentioned by several authors (Bensa, 1995; Bensa et al., 1982; Rivierre & Bensa, 1983), described by Jean-Marie Tjibaou as 'genealogies in space' (Tjibaou, 1996) that retrace the paths of the clans and make their history and migration physically visible (Levacher et al., 2017). As we progressed in our research, we realised that when we talked about the quality of the water, the Kanak people we spoke to always evoked the history of the water and therefore that of the clans. In fact, they explain that according to genealogy, the first group of the large clan comes to settle at the spring, where they begin to build the huts, start planting yams and taros, and when they have settled there, another part of the group splits off to continue its journey. They continue to follow the stream. Many tales on this subject can be found in Kanak literature, not only in the Paicî-Cèmuhi region; they can also be found in the Xârâcùù area, such as the story of the conch and the notou from the Thio region, collected by Arthur Maramin and transcribed by Yvon Kona, available in the archives of the Tjibaou Cultural Center. These stories trace the journeys of clans or the connections that water weaves between different clans, often in the form of humans, spirits, or animals. The narrative recounts the paths they have taken and the networks of alliances they have established along the water route (Grochain, 2020).

In the Hienghène region, as confirmed by Emmanuel Tjibaou, there is a clan attached to a specific water point and an eel. Water, therefore, serves as a reference point for determining land rights, the areas of influence of a group, just like peaks and ridge lines.

9.3.2.2 A Name, a Land, a River

> I mean you have your family name, your name, it refers to a specific place that was changed during the forced exile by the colonial administration. our name gives you a link to the land, it gives you a part where you can build your house, but always in adequacy with a creek that passes near by. Because the waterholes there are some people who take the names of the place that is the place of the ancestor, he is sitting there. And then they call the creek and the waterhole, and then they give them the name (Interview with Kokingone customaries, Touho, July 2022)

Today, many families like his, remain in places that do not correspond to the migration that follows the river but to a forced migration, an exile, orchestrated by the colonial administration in the nineteenth and twentieth centuries. Two stories, two narratives, therefore overlap on the landscape, complementing each other in order to understand the history of clan distribution across the territory.

9.3.2.3 Distribution of Roles Along the River

In the Cèmuhî genealogical narrative of the Galahi clan and the Waka clan from the Tiwaka and Kokingone tribes, recounted by two customary elders of the Kokingone tribe, they establish three groups: the one that stays up in the forest, the middle group and the group at the mouth of the river. It should be noted that the totems are also distributed along the watercourse.

> This other group, the last group to arrive at the mouth of the river, becomes the sea clan. And it becomes the patron of sharks, turtles, rays, sea cows and whales, since whales are also our indicator for starting the yam fields. (Interview with Kokingone customaries, Touho, July 2022)

In Hienghène, the mountain clans are indeed linked to totems of the mountain. For instance, on the ridgepole of the region, one can find the notou, a bird native to Kanaky/New Caledonia. Emmanuel Tjibaou draws a parallel between the call of the notou and that of the conch to reaffirm the continuous connection between the mountain clans and those of the sea.

My interviewees recount how the mountain and sea clans come together to exchange food:

> Once the clans are settled, they meet at the middle clans to engage in bartering; it's easier for them to descend, and it's easier for them to ascend. In this distribution, the Baï, the Galahi, are situated higher up, representing the forest clan, and the Waka, the Dui, are down below, representing the sea clan. But they always follow the course of the water. (Interview with Kokingone customaries, Touho, July 2022)

These places still exist and are called "Janap" in the Hienghène region and "Jenep" in Cèmuhî. One can still see the name "Jenep" at the communal market in the village of Touho.

9.4 Discussion: Troubled Waters

9.4.1 When Turbidity Means Danger

In the sections above, we have developed the idea that the notion of quality as proposed by the institutions does not allow us to account for the complexity of the links to water on customary lands. Starting with the term 'good water', we have shown that on customary lands, good water reflects the intimate individual and collective link between humans and their environment. If good water is synonymous with life and vitality, our interviewees also spoke to us about its opposite, turbid water, which is synonymous with tension and danger.

> And even you see, there are expressions when you say the most... Like now! Politics is unstable. Well, people say that this period is like in the Nèmi language, they say, We pulaak, it's the water that's cloudy. If you put your foot down to cross the river, you don't know whether you'll fall on a stone or a branch, you'll fall into the water, it's murky. Well, because when you say that water is cloudy, it's linked to life. If the water is cloudy, it means that your life is cloudy. It means that your foot on which you are standing, the ground on which you are leaning, on which you are going to walk, when you cross, it is hard, it means stable, it means something which is built, if on the contrary it is cloudy, it means that the spirits, there are things which are moving, fluids, dynamics which are not posed, therefore it is necessary to be careful. (Interview with Emmanuel Tjibaou, Tjibaou Centre, April 2022)

In Emmanuel Tjibaou's discourse, murky water is the translation of a problem, of instability in people's social life. Like a person crossing a murky river without being able to make out the obstacles on his path, the current social context does not allow for the future to be deciphered. We are in a period where murky water reflects strained social relations. Emmanuel Tjibaou refers to the third referendum on self-determination that took place in December 2021, the result of which is disputed by the Kanak population. In April 2022, we started the transition period that brings together politicians to draft a new status for the territory, but since the boycott of the third referendum, the pro-independence movement has refused any negotiation with the government. This "unstable" situation, represented by cloudy water, reveals to us the intimate links that underlie the relationship between water and humans. It reflects the quality or rather the non-quality links of people and their conflicts.

9.4.2 *Black Mulet: An Indicator of the River's Good Quality*

In the black mullet report, which looks at the reasons for the disappearance of the black mullet in the territory, particularly in the Thio and Ponérihouen regions, several people express the idea that the black mullet is a guarantor of good water quality because it likes clear, flowing water, as opposed to stagnant or poorly filled, murky water (see Chap. 8). The people involved in this project attribute the cloudy water and the lack of water to a lack of interest on the part of the population, which no longer cares for the rivers and creeks. This lack of interest is reflected in the social relationships (Autunno et al., 2022).

9.4.3 *Murky Water Reflecting Society's Discontinuous Flow*

> The cyclone, the flood, these are the reasons for war, when the water is dirty. As soon as the water rises, as soon as the quality of the water starts to be turbid, or when there's a current, that's synonymous with anger too. It refers to this idea of the spirits that may have sent you messages that you didn't read, so to make you react, well, it opens the floodgates… Then you are obliged to react. (Interview with Emmanuel Tjibaou, Tjibaou Centre, April 2022)

This interview adds the spirits in the relationship of water to humans who use the natural elements to send messages to humans. Water that rises due to bad weather, water that becomes dirty and dangerous, water that reflects war and anger according to Emmanuel Tjibaou. When there is a problem, it is not that the water is just dirty, it is that there is a problem with other humans or spirits. Dirty water signifies a dysfunction in society, a moment of tension, strong emotions that reflect society.

9.5 Conclusion

We wanted to show in this chapter that in Kanak cosmology the quality of water cannot be detached from its human and non-human environment, as it represents the link that humans have individually and collectively with water and their environment. If good water reflects the good functioning of society and, its harmony, we have seen that turbid water reflects the evil side, the ruptures of a society. Therefore, it is not sufficient to treat dirty water merely through chemical means. It is crucial not to overlook the social connections people have with each other, which are reflected in their relationship with water and contribute to ensuring the maintenance of good water quality.

These links must be taken into account and cared for, just like the members of the customary senate at the 2022 Water Forum in Noumea reminded us when they told us that debates on water must be directed towards social inequalities, because water is life. In the context of improving water management, this chapter emphasises the existence of water-related knowledge that remains underutilised or inadequately incorporated in institutional communication. Acknowledging and integrating these perspectives could undoubtedly enhance the inclusion and active participation of diverse populations in water management across the Kanaky/New Caledonia territory.

References

Anthonj, C., Tracy, J. W., Fleming, L., Shields, K. F., Tikoisuva, W. M., Kelly, E., Thakkar, M. B., Cronk, R., Overmars, M., & Bartram, J. (2020). Geographical inequalities in drinking water in the Solomon Islands. *Science of the Total Environment, 712*, 135241. https://doi.org/10.1016/j.scitotenv.2019.135241

Autunno, M., Bouard, S., Coulange, D., Le Meur, P.-Y., Richard, D., Despinoy, M., & Sabinot, C. (2022). *Projet MULET NOIR Pour une gestion intégrée et inclusive du milieu ripicole en Nouvelle-Calédonie: Une étude des représentations, des valeurs et des usages associés au mulet noir sur la côte Est de la Grande Terre* (p. 84). CNRT Nickel et son Environnement.

Bensa, A. (1995). Chroniques kanak. In *Ethnies (Paris 1985)* (pp. 18–19). Peuples Autochtones et Développement.

Bensa, A., Rivierre, J.-C., & Bensa, H. (1982). *Les Chemins de l'alliance: L'organisation sociale et ses représentations en Nouvelle-Calédonie (région de Touho—Aire linguistique Cèmuhi)*. SÉLAF.

Bouard, S., & Sourisseau, J.-M. (2010). Stratégies des ménages kanak: Entre identification sociale et abstraction marchande. *Natures Sciences Sociétés, 18*(3), 266–275.

Bouard, S., Lejars, C., Sabinot, C., Coulange, D., Grochain, S., Godin, P., David, V., Meur, P.-Y. L., Dotte, A.-L., Bearune, S., Ferrand, N., Despinoy, M., & Drouin, J. (2020). *Goutte (volet 2): La gouvernance de l'eau sur terres coutumières en Nouvelle-Calédonie: De la connaissance des usages et des savoirs à la production d'outils de gestion et de politiques adaptés* (p. 10). PUNC.

Bretteville, D. (2018). « Je ne sais pas ce que c'est, la coutume »: La coutume kanak par-delà tradition et modernité. *Journal de la Société des Océanistes, 2*, 473–486. https://doi.org/10.4000/jso.9502

Carpenter, C., & Lawedrau, A. (2002). Effects of forestry activities on surface water quality in the Pacific region: A case study of the Rewa River catchment, Fiji Islands. *International Forestry Review, 4*(4), 307–309. https://doi.org/10.1505/ifor.4.4.307.40536

Coulange, D. (2020). *Une étude de cas des vallées de la Pouanlotch et du haut de la Hienghène* (p. 110). CRESICA.

Coulange, D., Bouard, S., Grochain, S., Lejars, C., & Sabinot, C. (2021). Analyser et comprendre les représentations, les savoirs et les pratiques autour de l'eau sur terres coutumières pour appuyer la mise en œuvre de la Politique de l'Eau Partagée (PEP). *Revue Juridique Politique et Économique de Nouvelle-Calédonie, 37*, 16.

Crampton, A., & Ragusa, A. T. (2016). Exploring perceptions and behaviors about drinking water in Australia and New Zealand: Is it risky to drink water, when and why? *Hydrology, 3*(1), Article 1. https://doi.org/10.3390/hydrology3010008

DASS-NC. (2020). *BAROMÈTRE EAU-SANTÉ 1ère édition, 2020—Résultats d'analyse en Nouvelle-Calédonie*. The Pacific Community.

Davar, G. d. l. N. C. (2019). *Schéma d'orientation pour une politique de l'eau partagée* (p. 71).

Godelier, M. (2010). Systèmes de parenté, formes de famille Quelques problèmes contemporains qui se posent en Europe occidentale et en Euro-Amérique. *La Revue Lacanienne, 8*(3), 37–48. https://doi.org/10.3917/lrl.103.0037

Gosset, G. (n.d.). *Étude des usages et pratiques liées à l'eau potable à Voh, Koné et Pouembout* (Vol. 55). IAC et. CIRAD.

Gosset, L., Sabinot, C., & Worliczek, E. (2019). *Quand cyclones, pluies et pollution interrogent les liens des Kanak à leurs rivières et participent au renouvellement des savoirs écologiques (Thio, Nouvelle-Calédonie)* (p. 26).

Grellier, O. (2019). *Projet GOUTTE: Gouvernance de l'eau en terres coutumières* (p. 44).

Grellier, O., Bouard, S., & Lejars, C. (2019). *La gestion de l'eau en terres autochtones: Regards croisés sur la littérature dans le Pacifique et en Nouvelle-Calédonie* (GOUTTE) (p. 44). IAC, CIRAD, IRD, UNC, CRESICA.

Grochain, S. (2020). *Les cinq eaux kanakes* (p. 138). IAC.

Hughey, K. F. D., Kerr, G. N., & Cullen, R. (2010). *Public perceptions of New Zealand's environment: 2010* (EOS ecology) (p. 115). Lincoln University.

Lejars, C., Bouard, S., Sabinot, C., & Nékiriaï, C. (2019). Quand « l'eau, c'est le lien »: Suivre l'évolution des réseaux d'eau pour éclairer les pratiques et les transformations sociales dans les tribus kanak. *Développement Durable et Territoires, 10*(3), 15704. https://doi.org/10.4000/developpementdurable.15704

Levacher, C., Herrenschmidt, J.-B., Le Meur, P.-Y., Demmer, C., & Bouard, S. (2017). *Négocier, évaluer, reconnaître la valeur des lieux en Nouvelle-Calédonie, volume I. Rapport scientifique final* (Tome Nickel et Société [Progamme NERVAL]) (p. 67). CNRT Nickel et son Environnement.

Moglia, M., Perez, P., & Burn, S. (2008). Urbanization and water development in the Pacific Islands. *Development, 51*(1), 49–55. https://doi.org/10.1057/palgrave.development.1100451

Monnerie, D. (2012). Objets cérémoniels, chaînes opératoires et élaboration des relations sociales (arama et hoot ma whaap, kanaky Nouvelle-Calédonie, Mélanésie). *Techniques & Culture, 58*(1), 122–141. https://doi.org/10.4000/tc.6497

Mosley, L. M., Sharp, D. S., & Singh, S. (2004). Effects of a tropical cyclone on the drinking-water quality of a remote Pacific Island. *Disasters, 28*(4), 405–417. https://doi.org/10.1111/j.0361-3666.2004.00266.x

Nekiriai, C. (n.d.). *La gestion de l'eau sur terres coutumières en Nouvelle-Calédonie* (p. 104).

Nelson, S., Thomas, J., Jenkins, A., Naivalu, K., Naivalulevu, T., Naivalulevu, V., Mailautoka, K., Anthony, S., Ravoka, M., Jupiter, S. D., Mangubhai, S., Horwitz, P., Abimbola, S., & Negin, J. (2022). Perceptions of drinking water access and quality in rural indigenous villages in Fiji. *Water Practice and Technology, 17*(3), 719–730. https://doi.org/10.2166/wpt.2022.022

Rivierre, J.-C. T., & Bensa, A. T. (1983). *Histoires canaques*. Conseil International de la Langue Française: EDICEF.

Robak, A. (2022). Perceived vs measured water supply service: Evidence from New Zealand. *International Journal of Water Resources Development, 38*(6), 938–963. https://doi.org/10.1080/07900627.2021.1940105

Roumasset, J., Burnett, K. M., & Balisacan, A. M. (2010). *Sustainability science for watershed landscapes*. Institute of Southeast Asian Studies.

SOPAC & GWP Consultants. (2007). *Integrated water resources management in Pacific Island Countries: A synopsis* (p. 46).

Tjibaou, J.-M. (1996). In A. Bensa & É. Wittersheim (Eds.), *La présence Kanak*. O. Jacob.

White, I., & Falkland, T. (2015). Integrated management of urban water supply and water quality in developing Pacific Island countries. In Q. Grafton, K. A. Daniell, C. Nauges, J.-D. Rinaudo, & N. W. W. Chan (Eds.), *Understanding and managing urban water in transition* (pp. 489–526). Springer. https://doi.org/10.1007/978-94-017-9801-3_23

World Health Organization & Regional Office for the Western Pacific. (2005). *Report workshop on drinking water quality standards and monitoring for Pacific island countries* (p. 55). WHO.

Part III
The Shared Water Policy and Its Implementation

Chapter 10
The Shared Water Policy: From Diagnosis to Implementation

Pearl Winchester, Geoffroy Wotling, Gérard Fallon, and Aude Arrighi

Abstract The New Caledonia's Shared Water Policy (*la Politique de l'Eau Partagée* or PEP) orientation plan is the result of a diagnosis shared with all the territory's water stakeholders and of a process of inter-community co-construction. This chapter presents the main stages and proposals of the country's first water policy, unanimously adopted by the Congress of New Caledonia in 2019. It also discusses the challenges and threats of the actual application of the Shared Water Policy.

Keywords Public policy · Law · Competences · Political sphere

10.1 Introduction

The New Caledonia's Shared Water Policy (*la Politique de l'Eau Partagée* or PEP) orientation plan is the result of a diagnosis shared with all the territory's water stakeholders and of a process of inter-community co-construction. The diagnosis started being implemented by the "Assises de l'Eau" (a national water conference) in 2008, where the sharing of competences and the multiplicity of water stakeholders was clearly identified as a barrier to the development of a global public policy capable of addressing water issues in a proper measure. Between 2008 and 2019, the diagnosis was completed with assessments highlighting many technical, economical, societal, organizational issues: the freshwater resources are unevenly distributed and fragile, the Caledonians are unequal in terms of drinking water supply, overconsumption affects many towns, the water resources are under threat from fire, invasive species, mining. The diagnosis also emphasized on the unrecognized indispensability of the water resources for the tourism and industry sectors, and the

P. Winchester (✉) · G. Wotling · G. Fallon · A. Arrighi
DAVAR, Directorate of Veterinary Food and Rural Affairs, Water Division, Nouméa, New Caledonia

C. Lejars, S. Bouard (eds.), *Water Policy in New Caledonia*, Global Issues in Water Policy 32, https://doi.org/10.1007/978-3-031-88196-1_10

need to plan and anticipate the risks and impacts: developments impact water and water impacts developments. To tackle the issues pointed out by this diagnosis, an action plan was designed from the propositions created by participative workshops, led during the ForumH2O and during six local meetings (see Chap. 12). These first propositions were rearranged, refined and precised by different working groups between July 2018 and January 2019. The final PEP action plan, voted by the Congress on March 19th 2019, aims through six strategic objectives and four transversal objectives, to reconcile and respond simultaneously to the three challenges: water for nature, water for health and water for the economy (see https://eau.nc for more information). The governance of the Shared Water Policy has been structured in parallel, with the implementation of the Mission InterServices de l'Eau (MISE), gathering the different institutions in dedicated workgroups, the creation of the Comité de l'Eau (Water Committee), the decision-making body for the Shared Water Policy, made up of elected officials who represent all the institutions forming the MISE, and the organization of the annual Forum de l'Eau. After the adoption of the Shared Water Policy, it had often been said by the local water services that the PEP is a great engine, but it is lacking fuel to run. PROTEGE, the 11th regional EDF played as a first shared "fund" to implement the actions voted, before the creation of the Fonds de Soutien pour la Politique de l'Eau Partagée (support fund for the Shared Water Policy) in November 2021. Nevertheless, the actual application of the Shared Water Policy is still facing difficult challenges and threats, many years after its vote, for different reasons, although mainly because of recent political events and the missed opportunity to achieve legal and cultural pluralism through the water law (see Chap. 5).

This chapter presents the main stages and proposals of what is the country's first water policy. It will be divided in four sections: (1) From diagnosis to implementation of the Shared Water Policy (2) An ambitious action plan (3) Towards the implementation of the Shared Water Policy (4) What future for the Shared Water Policy? Strengths, weaknesses, threats but also opportunities, in particular essential debates on the status of water. After a reminder of the main stages and proposals of the water orientation plan, we finally discuss why its unanimous adoption by the Congress of New Caledonia in 2019 is not enough to guarantee its short-term implementation in a particularly complicated economic and political context. In a context where the political sphere is preoccupied with the challenges of the country's institutional future, where public financial accounts are stretched to the limit and in structural deficit, and where local authorities are narrowing their focus on their mandated tasks, it is natural to question the short-term future of a new country-wide policy, despite its acknowledged importance to all parties.

10.2 From Diagnosis to Shared Water Policy

10.2.1 First Steps Towards a Shared Water Policy

Water is a global sustainable development issue that is strongly impacted by climate change. In 2015, the United Nations set goal number 6: "Ensure access to water and sanitation for all and ensure sustainable management of water resources." This issue was addressed at the regional level by the Pacific Islands Forum (PIF), reaffirmed at the Pacific Water Conference ministerial meeting in Nouméa in 2018, and incorporated into the regional initiatives of the 11th European Development Fund (EDF).

To meet this challenge, the Shared Water Policy had to be integrated into the New Caledonia 2025 plan, the Do Kamo health plan, the energy transition plan, as well as provincial policies related to the environment, agriculture, mining and industrial development, urban planning and development plans for municipalities, as well as the actions of public institutions related to water issues, such as the Rural Agency, the Conservatory of Natural Spaces, the French Office of Biodiversity, or the Nickel Fund.

As early as 2008, the water conferences (*Assises de l'Eau*) organised by the Economic and Social Council (*Conseil Economique et Social*, CES) had identified the sharing of competences and the diversity of water stakeholders as a barrier to the development of a comprehensive public policy capable of effectively addressing water issues. Indeed, since the provincialisation and organic law of 1999, the administrative competences for water and aquatic environment management have been divided between four different authorities, lacking specific delineation, dedicated tools or means of coordination.

- New Caledonia is responsible for managing the public river domain (DPF) and is competent in matters of public hygiene, health, and civil security, including the prevention and management of natural risks.
- Customary authorities manage water resources that are not part of the public domain.
- Provinces, whose environmental competence includes aquatic environments, and which until recently for the North and South, exercised, by delegation, the management of withdrawal authorisations and the maintenance of watercourses.
- Municipalities (or their groupings), which manage the small water cycle, including the supply of drinking water and the treatment of wastewater; The French government may intervene in the event of a mayor's inadequacy.

These *Assises de l'Eau* clearly concluded the need for inter-departmental coordination, a clarified legal framework, and the establishment of local participatory management structures coordinated and supported by a Country structure. Until 2018, due to a lack of political support, its proposals had remained mere aspirations.

It was only in December 2017, following a series of droughts and water shortages that were hampering the country's development, particularly for agriculture, that the 15th government of New Caledonia decided to charge one of its members with building a water policy. Nicolas Metzdorf, a member of the government in charge of water and agriculture policy, supported by Didier Poidyaliwane in charge of ecology and customary affairs, and Valentine Eurisouke in charge of health, coordinated a broad consultation process involving various economic stakeholder, associations, and customary representatives to establish a framework for public policy orientation for water in New Caledonia.

Full details this participatory approach, which took place from February 2018, are provided in Chap. 12, but we can say that the participatory approach aimed, on the one hand, to develop a common and global vision of the issues, and on the other hand, to proactively involve stakeholders in the research and implementation of solutions.

10.2.2 A Shared Diagnosis

In 2018, the sharing of data between administrations, exchanges between technicians, customary officials, associations and elected officials have made it possible to collectively draw up many observations on the state of water across the country, and for the first time form a shared diagnosis of the freshwater situation in New-Caledonia. This diagnosis was written and shared through the Strategic Water Framework voted in 2019 by the Congress. The main outlines (DAVAR, 2018) are summed up in this section.

10.2.2.1 Unevenly Distributed and Fragile Water Resources

In New Caledonia, precipitation is very uneven in space and time. Between two and five times more water falls on the East coast than on the West coast and 50% of inflow is concentrated in the first 3 months of the year. In a context where natural reservoirs (i.e. watersheds) are of limited size and where the exploited resources are essentially superficial, these disparities result in significant vulnerability of New Caledonians to climate-related risks (for more details, see Chap. 2).

Paradoxically, the uses of water are in opposition to the available resources: "we need water the most where there is little, when there is the least!" The development of the country inevitably leads to an increase in vulnerability to droughts and floods, as well as an intensification of pressures between different economic, domestic and environmental uses, particularly in mining and agricultural regions (for more details, see Chap. 3).

10.2.2.2 Caledonians Unequal in Terms of Drinking Water Supply

If the inhabitants of Noumea, the capital, have secure water in both quality and quantity, the data available in 2018 evaluated that 5% of New Caledonians live without running water and 10% without drinking water. Field studies led by the government, through its Direction des Affaires Sanitaires et Sociales (DASS), are working towards a better qualification of these numbers. Also, more than 30% of distribution units have no treatment. Small communities are locked in a "vicious" cycle; the quality of service not being assured (red water, cuts, shortages, etc.), the bill recovery rate is low and overconsumption is enormous, which further reduces the means for network maintenance and improvement of services.

These inequalities in service provision are also reflected in the costs of drinking water ranging from 80 to 150 F/m^3, up to more than 800 F/m^3 when sea water must be desalinated. Furthermore, more than 90% watersheds supplying populations with water are in a degraded state and more than 40% do not even have the regulatory protection provided by water protection perimeters (PPE), which is likely to further exacerbate these inequalities.

10.2.2.3 Water Resources Under Threat

Fires and invasive species undermine the recharge of natural water reservoirs. The alteration of forests leads to the disappearance of soils, increasing phenomena such as runoff and erosion, thereby reducing infiltration into aquifers (see Chap. 3).

Furthermore, the altered peridotite mining sites affected by the almost exponential development of mining activity makes the water reservoirs particularly vulnerable. Nearly a third of the catchments are located on mining concessions and despite progress in water management practices, the effects of extractive activity on erosion and water turbidity are far from negligible.

When more than 30% of the surface area of a strategic resource such as that of Greater Nouméa with the Grand Tuyau on the Tontouta watershed allocated to mining companies, it prompts questions about the balance between the exploitation of green gold and the long-term preservation of the renewable nature of the water resource already qualified as blue gold.

On the other hand, the absence of a wastewater sanitation policy, the non-regulation of discharges outside of environmentally classified installations or even the intensification of agriculture are all threats which weigh more and more on New-Caledonia's vital resources.

10.2.2.4 Water, a Factor of Economic Wealth

In addition to these health and environmental issues, water also appears to be a crucial issue for the country's food self-sufficiency (see Chap. 3). Considering that most of the agricultural production areas are located in the driest sectors of the West coast, water appears to be a limiting factor requiring 20–30 million additional m^3 per year to reach 30% self-sufficiency by 2025, according to the province's agricultural policy (PPAP reference). While water needs to be made available to agriculture (by storing or transferring water), agriculture must also adapt to the available resource, by acting on production systems (choice of cultivated varieties, cultural practices, etc.).

Water is also an essential factor in the country's energy production, because the energy mix proposed by the Energy Transition Scheme, voted in 2016 (Deliberation No. 135 of June 23, 2016), provides for the development of hydroelectricity to reach 40 MW. This cannot be done without better control of the impacts of these levies on other uses and social acceptance of this type of development. Likewise, considerable volumes of water are also essential for metallurgy and industrial processes of thermal energy production.

Although too often ignored, water also represents a crucial issue for tourism development; hoteliers on the Isle of Pines and in the far North are well aware that without water, it is impossible to welcome guests. Furthermore, swimming sites, waterfalls and other aquatic heritage environments represent significant tourism potential that is largely undervalued.

10.2.2.5 Spaces and People Under the Transformative Influence of Water

Developments impact water and water impacts developments. Nouméa was built on a peninsula without water, and the Voh-Koné-Pouembout development zone has for many years been the focal point of social unrest linked to water shortages, competing uses and poor water quality of the water distributed.[1]

Planning and anticipation of risks and impacts is often insufficient and a source of tension, conflict or even drama (Houaïlou, Kouaoua, Thio, Poya, etc.). Current means of prevention and protection of property and people do not allow them to respond to the challenges. For example, in 2018, more than 13,000 constructions were recorded in flood zones.

To respond to the threats, the needs already identified and the challenges of climate change, the investments available for developments such as regulatory tools like "Avoid-Reduce-Compensate" remain largely insufficient in New Caledonia. Preserving, restoring natural reservoirs, maintaining, developing artificial reservoirs, reducing pollution at the source, treating it before discharge, controlling and limiting withdrawals such as effluents, require much more powerful economic and regulatory levers than those existing at the national level at the time of this diagnosis.

[1] Demonstrations and arson against the SIVOM VKP, establishment of the VKP Water Management Committee to re-establish dialogue and manage shortages and conflicts of use, delays in the opening of the Northern Hospital due to poor quality of water distributed.

10.3 An Ambitious Strategic Water Framework and Action Plan

The Shared Water Policy strategic framework (Gouvernement de NC, 2019) and the action plan linked to the framework were created from the propositions put forward at the workshops during the ForumH2O and received during the local meetings in Boulouparis, Kaala-Gomen, Hienghène, Houaïlou, Lifou. These first propositions were rearranged, refined and detailed by different working groups between July 2018 and January 2019 (see Chap. 12 for more details). The final PEP diagnosis and action plan were drafted by many administration officers and water stakeholders, under the coordination of Mr Gérard Fallon, then director of the DAVAR.

Building on these widely shared findings and in order to restore, consolidate or perpetuate the balance between water, Caledonians and their environment, the PEP framework and action plan has six strategic objectives that aim to reconcile and respond simultaneously to the three challenges: water for nature, water for health and water for the economy.

Strategic Objective 1: Protect catchment areas and strategic resources, preserve environments

All stakeholders agree that our first responsibility is to safeguard the sustainability of the exceptional aquatic environments and the availability of water resources, in sufficient quantity and quality, so as not to compromise the development of future generations.

It has been suggested that so-called "strategic" resources be defined, meaning those whose irreversible degradation poses a significant threat to vital water uses for health, the economy, or the environment, in order to focus the means of preservation there, and to ban fires, invasive species and activities likely to compromise their integrity.

Strategic Objective 2: Provide 150 L of drinking water per day per Caledonian

Water is a "health determinant" of the Do Kamo plan, the Caledonian health plan voted in 2018 (Deliberation No. 365 of 19 December 2018). 150 L per day per Caledonian is a realistic objective but one which will have to adapt to the diversity of local contexts. This objective involves implementing Water Health Safety Plans, revising drinking water standards, improving distribution networks and combating overconsumption. It must also allow the exploitation of alternative resources in the event of insufficiency or contamination of existing resources and also guarantee access to drinking water without health risks in times of crisis.

Strategic Objective 3: Make all Caledonians hydro-eco-citizens

Instilling the importance of water and the environment in the consciousness of all New Caledonians is the societal foundation the Shared Water Policy needs to succeed. This awareness was strongly affirmed during the H2O Forum, and needs to be shared with all New Caledonian society. It is in fact a question of recreating the link between tap water and the state of the watershed, between toilet water and the state of the lagoon. Information-awareness campaigns must be carried out, of course in schools, but also with businesses, customary officials, municipal councils, etc.

Strategic Objective 4: Move towards zero discharge of untreated water

Improving New-Caledonia's wastewater management and combating chronic or accidental pollution are essential objectives for safeguarding the health of the country's inhabitants and preserving the state of New-Caledonia's rivers, lagoon and reefs. The effort made for collective sanitation must be continued but above all, the performance of non-collective sanitation must be improved in areas with low housing density, and must be accompanied by a definition of discharge standards, and appropriate control. Regulation according to the "polluter pays" principle of diffuse pollution, particularly of agricultural or mining origin, is of course included in this strategic objective.

Strategic Objective 5: Put water at the centre of planning, housing and economic development projects

Anticipating a crisis, planning according to water and for water are objectives which aim to promote sustainable development, adapted to the environment. Development projects must be planned and integrated in different forms: flood-prone area development plans, drinking water and sanitation master plans, etc. They must also be part of a risk prevention and pollution approach according to the "Avoid-Reduce-Compensate" concept.

Strategic Objective 6: Better control of freshwater to increase local agricultural production

To guarantee their food, Caledonians must promote agriculture while controlling water consumption and its impacts. To reconcile these imperatives, two avenues are possible: adapting agriculture to water and adapting water to agriculture. Adapting agriculture requires a change in practices and consistency in agricultural policies; adapting water requires significant investments in infrastructure projects.

These six pillars of the PEP are supported by **four transversal objectives**, supporting actions which make it possible to respond to several strategic objectives in a coherent manner.

Transversal Objective A: Establish a legal framework and a policy adapted to an effective water policy

Legal support is a prerequisite for achieving all water policy objectives. To establish water law, the definition of a water status for the general interest and dissociated from land status appeared essential, and extensive legal work must be carried out, particularly with regard to standards for drinking water, discharges, withdrawals and management of the public river domain.

Transversal Objective B: Improve knowledge to better protect, preserve, plan and manage

Knowledge is the basis of the water policy dashboard. It is essential for management, forecasting and awareness. As a priority, sharing information and data relating to water through standard formats and an interoperable platform have been identified as the first avenues for progress. No governance without knowledge, no management without indicators. Relevant, up-to-date and shared information is therefore needed.

Transversal Objective C: Develop the skills of all water stakeholders for better performance

Skills development is also essential to achieving the Water Policy objectives. Water professions are very specific and the needs concern both the public and private spheres. Skills mapping and networking expertise is a priority action in the short term to develop mentoring and local skills before mobilising international training tools.

Transversal Objective D: Water Governance and Financing

The orientation plan will truly become a public policy once it has a governance body and an economic model that aligns with the objectives adopted. Within the constraints of the political agenda, these subjects, being more political than technical in nature, have not been thoroughly addressed. It was firstly a matter of securing consensus on what needs to be done, before trying to figure out how it should be done. Nonetheless, the imbalances between the small and large water cycles have been clearly identified, as has the lack of coordination among the diverse stakeholders involved in the field of PEP.

10.4 Toward the Implementation of the Shared Water Policy

Without a pilot and without an engine, it is naturally difficult to progress, however, the PEP benefited in 2019 and 2020 from the commitment of the inter-departmental mission, the support of Jean-Pierre Djaiwe, member of the government chairing the water and resources committee of the 11th European Development Fund.

10.4.1 The Organisation of the Mission Inter-services de l'Eau (MISE), the Water Committee (Comité de l'Eau) and the Forum de l'Eau

Following the adoption of the plan, the *Mission Inter-services de l'Eau* (MISE) was structured according to a logical framework, with a working group assigned to Strategic or Transversal Objective, i.e. ten different working groups in total. These working groups are inter-institutional, including representatives from all communities (Fig. 10.1).

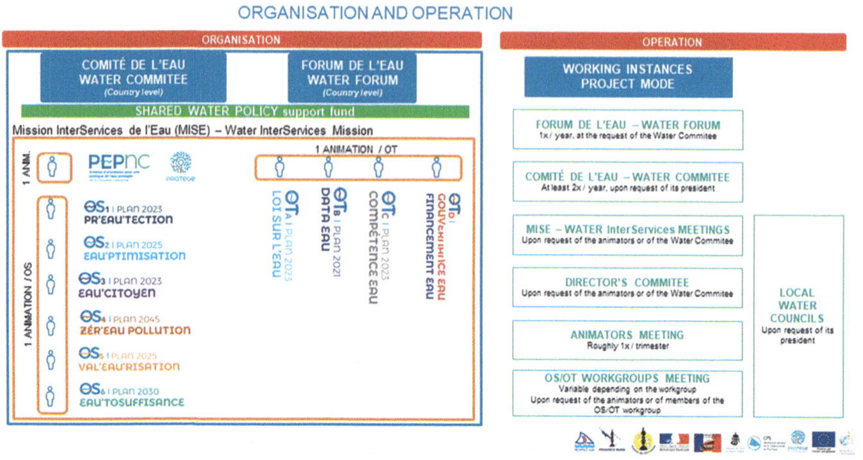

Fig. 10.1 Governance and organisation of the Mission InterServices de l'Eau

Working group leaders gather at least once a quarter during a collegial meeting coordinated by the Water Department of the Government of New-Caledonia.[2] This meeting allows facilitators to discuss the actions carried out and cross-cutting topics for several groups. It is also an opportunity to identify difficulties in the implementation of actions and in particular the lack of support on certain topics, to prepare the water committees and the annual forum. The coordination of these meetings also facilitates updates to the "eau.nc" website and the distribution of a monthly newsletter.

The operational MISE was officially launched in June 2019, with the gathering of more than 50 administrative officers from all the institutions of New Caledonia. The working groups were implemented between July and September 2019, with their leaders appointed before the end of September 2019. At roughly the same time, and officially on 20 November 2019, the main coordinator was recruited for at least 3 years (2019–2022), made possible through European financing via the PROTEGE program.

The role of this main coordinator of the Shared Water Policy was to keep the network of working groups alive and running, coordinate the actions between all the institutions and coordinate the actions financed by Europe with the PROTEGE program. If needed, the main coordinator would be the project manager for some of the PROTEGE actions and projects, to ensure their effective implementation.

At first every working group was composed of at least three people, and a coordinator was elected or appointed for each of them. Except for the OS4 group, which was coordinated by a Province Nord official together with the Director of the French

[2] Service de l'Eau du gouvernement de la Nouvelle-Calédonie.

Mayor's Association,[3] and for the OTC group, which was coordinated by DAFE[4] representatives, every appointed coordinator was an official of the government of New Caledonia. The Province Sud, the province des Iles, the Senat coutumier and the communes did not put forward any coordinators to lead a working group.

Before the end of December 2019, many groups' coordinators had stepped down and no-one else had taken on the role. This is partly because the coordination tasks are additional to those already assigned to the candidate, with no compensation being offered meaning that there was no dedicated time allocated and no financial incentive for the new responsibility.

By December 2023, four of the ten working groups did not have coordinators, and the six other groups were coordinated by a New-Caledonia government official. Needless to say that the sharing part of the Shared Water Policy is difficult to implement between the different institutions. Nevertheless, the Water Department does its best, with the financial help of Europe through PROTEGE, to keep the MISE active, by organising the coordinators' meeting every trimester.

10.4.2 Comité de l'Eau: The Water Committee

The *Comité de l'Eau* is the decision-making body for the Shared Water Policy. It is made up of elected officials who represent the institutions that form the MISE:

Members of government in charge of water policy, agriculture, sustainable development, customary affairs, health
The presidents of provincial assemblies
The president of the customary senate
The presidents of the two mayors' associations

It was established on 20 February 2019 by Deliberation No. 395 of the Congress,[5] at the same time as the *Forum de l'Eau*, and its composition was changed in September 2023 (see "The support fund for a shared water policy").

The *Comité de l'Eau*, chaired by the government and bringing together representatives of the different provincial communities, mayors' associations and the customary senate, is the leading and deciding committee for the strategic direction of the Freshwater Policy.

It first met on 2 October 2019 to validate the use of a participatory approach to the implementation of the scheme. With the support of the IAC, the CIRAD and IINRAE (G-EAU[6] Research Unit), innovative consultation methods were tested in

[3] Association Française des Maires AFM.

[4] DAFE: Direction de l'Agriculture, de la Forêt et de l'Environnement, Haut-Commissariat de la République.

[5] https://juridoc.gouv.nc/juridoc/jdwebe.nsf/joncentry?openpage&ap=2019&page=2372

[6] http://www.g-eau.fr/index.php/fr/umr-geau

three pilot regions (Hienghène, La Foa-Boulouparis and Maré) in order to prioritise and adapt the PEP action plan according to the typology of the territories.

In 2020, following the COVID19 health crisis, the *Comité de l'Eau* only met once to nominate a facilitator to the governance working group (OTD) and decided not to hold a forum at the end of the year but to schedule the 2021 edition at the start of the year, on World Water Day (March 23). This was without taking into account the fall of the government (2 February 2021) and the second and third COVID19 lockdowns (March 2021 and September 2021). Only one *Comité de l'Eau* meeting could be held that year, on 21 December 2021.

With the election of Joseph Manaute as the member of the 17th government of New Caledonia in charge of the Shared Freshwater policy in February 2021, two meetings of the *Comité de l'Eau* were organised in 2022. Mr Manaute has been at the heart of the creation of the Support Fund for a Shared Water Policy, adopted by the Congress on 5 November 2021 (Deliberation 50/CP[7]) (see "The support fund for a shared water policy), for which a special decision-making committee was established.

In December 2022, after the resignation of Mr Manaute, Mr Jérémie Katidjo-Monnier became the member of the government in charge of the Shared Water Policy. As president of the *Comité de l'Eau*, he organised one meeting of this body in 2023.

10.4.3 Forum de l'Eau

On 26 November 2019, the results of the workshops coordinated by the IAC and the G-EAU unit of CIRAD in the three pilot regions were presented to more than 200 people during the *Forum de l'Eau* 2019.[8] The water stakeholders who attended also discussed priorities and implementation solutions, particularly in multi-stakeholder groups of 3–5 people.

No Forums could be organised in 2020 and 2021, mainly because of the COVID19 crisis and the budget crisis.

In 2022, a second *Forum de l'Eau* was organised in March, under the direction of Joseph Manaute, member of the 17th government of New-Caledonia in charge of the Freshwater Shared Policy. Again a success, with the gathering of more than 180 water stakeholders over 3 days, this second *Forum de l'Eau* was focused around co-construction and participatory workshops, like the Forum H2O in 2018 and the first *Forum de l'Eau* in 2019. This time, the topics of discussion revolved around Local Water Councils (*Conseils locaux de l'Eau*), governance and financing priorities amid restricted timeframes and public budgets.

[7] https://juridoc.gouv.nc/juridoc/jdwebe.nsf/joncentry?openpage&ap=2021&page=16587

[8] https://eau.nc/forum-de-leau-2019

Joseph Manaute was replaced by Jérémie Katidjo-Monnier in December 2022 and no *Forum de l'Eau* could be organised in 2023 in New-Caledonia, due to budget restrictions.

10.4.4 The 11th Regional EDF, a First "Shared" Fund

While there was no specific budget associated with the adoption of the plan, the MISE was able to receive funding from the "WATER" programming of PROTEGE[9] (Oceanian Regional Project of Territories for the Sustainable Management of Ecosystems, 11th European Development Fund) for the initial implementation of the PEP action plan. Indeed, this European program provides 4.320 billion XPF (36 million €) to the four OCTs, including around 300 million XPF (2.5 million €) to New Caledonia on the "Water" theme.

As a result, from 2019, the MISE working groups were used to establish specifications and launch PROTEGE calls for projects, in line with the PEP action plan. This includes, for example, for OS1, watershed restoration actions aimed at influencing the major water cycle, for OS2, the identification of isolated areas and the establishment of water purification stations, for OS4, the study of a sludge treatment and wastewater reuse sector, for OS5, a project to improve resilience to flooding and for OS6 the installation of connected meters for irrigation.

The PEP was able to become a reality through the opportunity offered by this "shared" fund, benefiting all local authorities and stakeholders through practical on-the-ground initiatives. Even if these are only pilots, with at best a symbolic impact on achieving the objectives of the plan, they demonstrate the interest of a dedicated fund, however limited it may be. The challenge now is to be able to sustain this dynamic in the long term by seeking continued access to these financing opportunities using a similar method capable of taking over from the 11th EDF which was planned to end in 2022.

Due to the COVID19 health crisis, the PROTEGE program (11th EDF) has been postponed to September 2024, and March 2024 for the Freshwater actions in New Caledonia. By September 2023, 90% of the water actions for New Caledonia planned in the program were over, with 75% of the budget for implementation utilised at that date. This program has been a huge success for Water in New Caledonia, helping all the institutions (government, provinces, communes, associations) to work together to reach the objectives, and a significant driver behind the implementation of the initial Freshwater Shared Policy actions.

[9] https://protege.spc.int/fr

10.4.5 The Support Fund for a Shared Water Policy

Under the leadership of Joseph Manaute, a support fund for the Shared Water Policy was set up on 5 November 2021, by Deliberation No. 50/CP of the Congress of New Caledonia. This fund, housed within the government of New Caledonia, uses revenue from the Agricultural Production Support Tax (TSPA) and the Tax on Insurance Agreements (TCA), up to 5% for each. Hosted by the Rural Agency, until 2021, this payment provides the Shared Water Policy with an annual budget of around 150 million XPF to implement actions.

This fund is administered by the Management Committee of the Support Fund for Shared Water Policy, which is made up of the following members:

- Member of the government in charge of water policy
- Four representatives of the Congrès de la Nouvelle-Calédonie
- The presidents of provincial assemblies
- The president of the CESE
- The president of the customary senate
- The presidents of the two mayors' associations
- The president of the Chambre d'Agriculture et de la Pêche
- Two qualified people, representing civil society

From the 15 members of the *Comité de gestion du Fonds de soutien* listed above, 7 are also members of the previously created *Comité de l'Eau.*

The *Comité de gestion du fonds de soutien* met twice in 2022, and once in 2023. In 2023, it was suggested that the *Comité de gestion du fonds de soutien* be merged with the *Comité de l'Eau.* This merger was approved by the members of both committees in September 2023.

10.5 What Future for the Shared Water Policy?

In a context where the political sphere is monopolised by the challenges of the country's institutional future, where public accounts are lifeless and in structural deficit, while communities each focus on their mandated tasks, it is natural to question the short-term future of a new country-wide policy, despite its acknowledged importance to all parties.

10.5.1 Advantages

Notwithstanding, the PEP offers a structured logical framework that allows individual and collective initiatives to combine their actions into a coherent whole. This support can legitimise the efforts of the different partners by more clearly identifying the interactions between the different projects. Through the water committee, established by Congress, the government now has a clearly identified member, responsible for leading the water sector and a space for interinstitutional dialogue on water-related issues. Likewise, the *Forum de l'Eau* sets an annual meeting with civil society and water stakeholders, which involves the publication of an annual report on the progress of the PEP, likely to sustain progress moving forward.

Also, the participatory approach which naturally imposed itself in the development and implementation process is aimed directly at voters, which allows the PEP to have a set of "supporters" capable of raising awareness among elected officials or administrative authorities to water issues.

Finally, in addition to the resources provided by the 11th EDF, the creation of the Support Fund for the shared water policy allows the government to have a minimum of allocated royalties (>5% TSPA, TCA) which although modest, 100–150 Million XPF/year (833 k€ to 1250 million €), constitute the first contributions to water policy support funds.

The existence of an approved plan also makes it easier to access support funds provided by national donors such as the OFB (French Biodiversity Office, merger of the French Biodiversity Agency and the French Biodiversity Office National Water and Aquatic Environments) or international ones such as the KIWA initiative supported by the AFD (French Development Agency).

Consequently, in 2022, a first call for tenders was opened by the government of New Caledonia via the Support Fund for the Shared Water Policy, for an amount of 50 million F (~416 k€), in order to support concrete actions carried out by local associations, businesses or communities. Fourteen projects were selected and financed in 2023. Building on this first experience, a second call for tenders was opened at the end of 2023 (see https://eau.nc). But financial support for water policy remains low given the stakes involved.

10.5.2 Weaknesses

It is of course the absence of a dedicated budget commensurate with the strategic objectives announced which constitutes one of the most important obstacles to the implementation of the PEP. This lack of specific resources causes concerns among various partners regarding requests for contributions from their own funds.

Also, the country dynamic which led to the development of the plan lost considerable momentum at the time of its implementation. The water committee had the greatest difficulty in mobilising provincial elected officials, work on PEP governance and financing methods has not really progressed, with certain key players having declined the invitation to dialogue. Furthermore, the comprehensive action plan lacks prioritisation. If, as the 2019 and 2022 forums demonstrated, the participatory approach can be successful in defining contextualised priorities, it is up to the different executives to consult together, to position themselves and to adapt budgets accordingly.

10.5.3 Threats

In the current economic situation, the political dynamic is not conducive to "Country" openness and solidarity, especially from the south province. Considering their insufficient level of remuneration, the provincial communities have opted to abandon the delegation of water resource management, which has been in place since provincialisation and regulated by Deliberation No. 238 of 1997.

As a result, this competence was returned to New Caledonia at the start of 2021 even though the government services have no new means to exercise it. This disengagement is also reflected in the MISE working groups, some of which remain orphans while waiting for a leader. It is evident that without a clear directive, community representatives are no longer willingly involved in the work as they were at the start of the PEP and those who remain motivated are exhausted by not being supported by their hierarchy and their executives.

10.5.4 Opportunities

However, while the abandoning of water management was not the scenario the MISE had hoped for, it effectively represents administrative simplification as included in the expected results of the PEP. This change will force rapid reforms in the organisation of services as well as in the mode of operational intervention.

While the administrative authority could indeed become centralised, operational management cannot be conducted from the offices of the capital. The participatory approach involving local actors and including multi-level governance starts to emerge as the only alternative for integrated management of water resources.

Furthermore, to face its new responsibilities, New Caledonia must urgently develop a national law to precisely define the duties and responsibilities of users and owners of the public river domain. This law was co-constructed between 2021 and 2023 through 13 participatory meetings, bringing together all New Caledonian institutions. As of January 2024, this draft law has been brought before the Congress and generated open debates on the status of water and the resources allocated to its proper management. But more than 8 months after its first presentation to the elected representatives of the Congress in 2023, the law on the public domain of freshwater has still not been adopted.

At the same time, the provinces abandoned their authority over water management, and have shown limited interest in working on governance and financing, despite the fact that economic development and environmental protection cannot be mutually exclusive and both rely heavily on water, especially in the agricultural sector.

Finally, the exit from the Nouméa agreement and the revision of the organic law which could begin, also constitutes an opportunity to finally establish a better distribution of authority relating to water management, particularly with regard to other resources, like natural resources and development and land-use planning skills.

10.6 Conclusion and Perspectives

In 2024, the political and economic situation in New-Caledonia is still not conducive to the launch of a country-wide orientation plan that deals essentially with long-term issues. The necessary budgetary arbitrations and parliamentary debates on the sharing of authority should have constituted an opportunity for progress in water management, but elected representatives are now more focused on New-Caledonia's status following the Accord de Nouméa.

However, thanks to the creation of the support fund for the Shared Water Policy and its associated *Comité de l'eau* (Water Management Committee), discussion and exchanges at the higher level of elected representatives started in 2022, and co-financing from external entities, such as the French Office of Biodiversity, was made possible. The proposed law on the water sector, which was co-constructed, also made it possible to raise awareness among elected representatives of the Congress about the difficulties of implementing the Shared Water Policy action plan. Finally, the launch of the two calls for tenders in 2022 and 2023 from the Support Fund for the Shared Water Policy have made it possible to push forward projects supported by stakeholders outside the MISE that fully align with the Shared Water Policy action plan. Nevertheless, the actual application of the Shared Water Policy is still facing difficult challenges and threats, many years after its vote, for different reasons, although mainly political.

References

DAVAR (Direction des Affaires Vétérinaires, Alimentaires et Rurales). (2018). *Forum H2O: Quelques chiffres clés, document présenté le 18,19 et 20 avril 2008 à Nouméa, disponible en ligne sur*. Retrieved from https://eau.nc/forum-h2o-2018 et sur https://davar.gouv.nc

EAU.NC. The Shared Water Policy website. Retrieved from https://eau.nc

Gouvernement de la Nouvelle-Calédonie. (2019). *Politique de l'eau partagée, schéma d'orientation pour une politique de l'eau partagée de la Nouvelle-Calédonie*. Retrieved from https://davar.gouv.nc/sites/default/files/atoms/files/pep__schema_dorientation_pour_une_politique_de_leau_partagee_de_la_nouvelle-caledonie.pdf.

Chapter 11
The VKP Water Management Committee: An Inspiring and Successful Experience for Water Management and Stakeholder Engagement

Marion Bois, Séverine Bouard, Caroline Lejars, and Nils Ferrand

Abstract Before the launch of the participatory process aimed at implementing a Shared Water Policy in 2018, several local consultations had been initiated at the beginning of the 2010s in catchment areas where concurrent water uses were becoming problematic. The very first experience of participation in New Caledonia took place in 2015 with the Voh-Koné-Pouembout local water management committee. This local experience was one of the cornerstones of the methodology used for the Shared Water Policy. This chapter recounts this shared experience. The municipalities of Voh-Koné- and Pouembout villages, located in the North-West of New Caledonia, faced unprecedented development generated by the establishment of an international scale nickel mining industrial complex, the Koniambo Nickel Project, from 2006 onwards. To cope with the risks of shortages, quality degradation and conflict over water rights, the Northern Province created a local water committee (the Water Management Committee of Voh-Koné-Pouembout—VKP WMC) to facilitate regional development while safeguarding water resources. The VKP WMC members (experts, customary leaders, farmers, etc.) have developed an integrated water resource management program using participatory methods. The authors, directly involved in this local participation process, describe the tools they

M. Bois (✉)
DAF, Planning and Land Management Directorate, Province Nord, Kone, New Caledonia
e-mail: m.bois@province-nord.nc

S. Bouard
IAC (New Caledonian Institute of Agronomy), Pouembout, New Caledonia

C. Lejars
CIRAD (French Agricultural Research Centre for International Development), UMR GEAU (Univ Montpellier, CIRAD, INRAE, IRD, BRGM, AgroParisTech, Institut Agro), Montpellier, France

N. Ferrand
INRAE, Research Institute for Agriculture, Food and Environment, UMR GEAU, Montpellier, France

C. Lejars, S. Bouard (eds.), *Water Policy in New Caledonia*, Global Issues in Water Policy 32, https://doi.org/10.1007/978-3-031-88196-1_11

used to facilitate water management coordination. This experience of governance was unique to New Caledonia and has been praised by all its members for its ability to unite and coordinate actions that benefit the population.

Keywords Participatory tools · Role playing-games · Dam project · Water management

Learning to share a resource can be a challenging task, especially when its fragile nature sometimes results in incompatible uses. New Caledonia's institutional complexity and high political stakes make it a dynamic testing ground for innovative approaches to resource management. In the Voh-Koné-Pouembout area, a water management committee was established to bring together stakeholders in an innovative format. New participatory approaches were also developed. This chapter presents the VKP Water Management Committee initiative, which was implemented in response to the challenges faced by the VKP zone. We will examine how public institutions and research bodies collaborated to provide training in territorial consultation and its application to water management. Finally, we will evaluate these experiences to determine if a culture of participation has been established in this context.

11.1 Facing the Growing Challenges of Water Resource Management: Dynamic Local Water Governance in VKP

11.1.1 A Need for Knowledge Sharing to Face Complex Issues Associated with Water Management

The rural region of the Voh-Koné-Pouembout communes (1900 km^2), located in the North-West of New Caledonia, has experienced unprecedented development, driven by the establishment and operations of the KNS mining complex, growth which has led to the doubling of the population in 10 years.

This economic rebalancing in favour of the North of the country, with its Kanak majority, is the result of the Matignon Accords—an agreement aimed at giving New Caledonians the means to develop economically, institutionally and culturally, before deciding on their independence from France.

This rapid development process has led to the expansion of all kinds of artisanal, industrial and agricultural activities in the VKP zone and has generated an increasing pressure on water resources.

Given the complex distribution of authority in the water sector, spread across multiple institutions, it was crucial to bring all stakeholders to the table to establish a shared understanding and coordinate actions toward a common goal: integrated water resource management in the context of a changing territory.

An original and informal structure bringing together all the institutional stakeholders and user representatives involved in water management was created in 2010: the VKP WMC. Initially, it was through the spontaneous initiative of different local water stakeholders and managers (technicians, officers from institutions involved in water issues). Later, the North Province, the competent authority for planning, economic development and the environment, led and organised the VKP committee and developed innovative planning tools to ensure that the VKP zone development also protected its water resources.

The advantage of this unique committee, created on a voluntary basis, was its capacity to consider local social and cultural nuances without being limited by a rigid and formal framework.

11.1.2 The VKP Water Management Committee: Adaptation to Issues and Ability to Engage Researchers in Developing Their Collective Analytical Skills

This committee brings together several stakeholders: the State, several directorate of the New Caledonian Government (Department of Health and Social Affairs (DASS), Directorate of Veterinary, Food and Rural Affairs (DAVAR), Directorate of Industry, Mines and Energy (DIMENC), Geological Survey of New Caledonia (SGNC)), several departments of the North Province (Department of Agriculture, *Service de l'aménagement et de la gestion de l'eau* (SAGE), *Direction du Développement Économique et de l'Environnement* (DDEE), Cellule Koniambo), the communes of Voh, Koné and Pouembout, SIVOM VKP (multi-purpose intermunicipal syndicate), Météo France Nouvelle Calédonie, Chamber of Agriculture, University of New Caledonia, Association of Agricultural Water Users of Pouembout (AUEA P) and an environmental association (AMU KEJE).

Gathered around the North Province, the members of the VKP WMC have drawn up an integrated water resource management program, including around a 100 actions over the first 3 years. The plan seeks to characterise the state of adequacy between needs and resources, risk management, preservation of the environment and heritage as well as development of water governance methods. After 2 years of studies, it was indisputable that there was an imbalance between needs and resources in both the current and future situations (up to 2050), based on available planning documents. The committee operates according to the initiative and means of its members and through the financial support of the European Commission and the South Pacific Commission. The North Province drives the dynamic implementation of the actions by coordinating the WMC and providing technical and financial support.

However, shortly after the first 3 years, the idea of setting up a participatory approach in the territories of Voh, Koné and Pouembout was tabled in 2014, with the aim of building a future shared program for the committee but also of developing specific tools to further share local issues related to water management. The same year saw the implementation of a study on "Water governance in the Pacific: uses, public policy mechanisms" as part of the "Freshwater management" project set up by the Pacific Fund[1]. This 18-month project was conducted by various actors: Australian National University (ANU) within a private-public consortium, Institut Agronomique Néo Calédonien (IAC), Institut National de Recherche en Sciences et Technologies pour l'Environnement et l'Agriculture (IRSTEA) and the Institut de Recherche pour le Développement (IRD). Faced with the increase in water-related problems in the Pacific zone and to compensate for the lack of research on these issues, particularly in social sciences, this project seeks to identify competition between water uses (agricultural, food and industrial) by analysing local water policies from a historical perspective.

To do so, it was structured around four research themes:

Identifying competition and conflicts around water uses.
Analysing local water policies from a historical perspective.
Describing water management processes and mechanisms, and stakeholders involved.
Understand how local water policies are influenced by other levels of governance.

Four case studies were comparatively analysed: Australia, Solomon Islands, Republic of Kiribati and New Caledonia.

In New Caledonia, the VKP area quickly attracted attention and became a study area: the establishment of the international metallurgical project in Voh brought about many changes through a rapid increase in activities and population, highlighting the competition for access to water. At the same time, there were many other pre-existing issues in the region, such as the introduction of water pricing and the emergence of environmental issues (climate change leading to changes in water status, e.g., the phenomenon of the rising salt water wedge). Moreover, many solutions and actions were discussed by local stakeholders and with VKP WMC'S input.

One key outcome was the progressive emergence of the idea to create a multipurpose dam, as local territorial institutions revived a project aimed at strengthening water resources (Fig. 11.1).

Multiple stakeholders and competition for water use, as well as issues relating to water quantity and quality, have therefore raised questions about the fair distribution of water resources in the local area. The extensive fertile plains of Pouembout are also a key area for agriculture, especially as New Caledonia seeks to be increasingly self-sufficient. Irrigation is expected to increase rapidly in the context of agricultural expansion and climate change. The first social sciences studies have since been conducted with local water representatives in Pouembout, including farmers, Kanak

[1] The Economic, Social and Cultural Cooperation Fund for the Pacific or "Pacific Fund" (PF) is one of France's regional cooperation instruments in the Pacific. The PF is financed by credits from the Ministry of Europe and Foreign Affairs (MEAE) and contributes to the regional integration of New Caledonia, French Polynesia and the Islands of Wallis and Futuna through cooperation with the independent States of the Pacific.

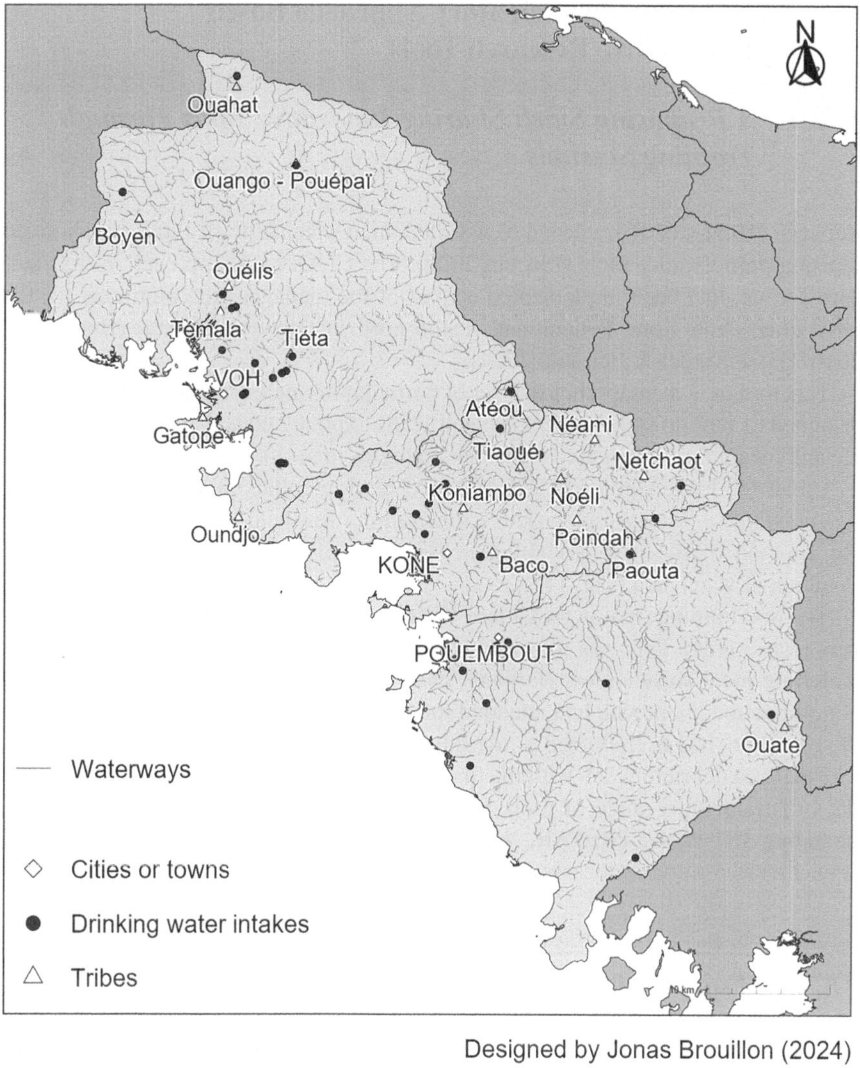

Fig. 11.1 Voh, Koné, Pouembout

people, village residents, etc., and a broader participatory approach has been introduced (Chevreux Warode, 2015). This was made possible by a strong local commitment from the stakeholders involved in the Water Management Committee and its objective to encourage consultation between the area's various water users. The research project then trained stakeholders in how to use a comprehensive set of tools to implement a participatory modelling system, aimed at facilitating the development of public policies in water management.

11.2 An Original Participatory Approach Based on Innovative Research Tools

11.2.1 A Promising Start: Sharing Innovative Tools Through Training Sessions

Between 2015 and 2017, a dozen New Caledonian water authority staff were trained in using participatory tools (Pizette, 2020). The COOPLAN approach, in its standard format, had been implemented as part of the management committee for the VKP zone (Voh Kone Pouembout, in the north of New Caledonia) (Bouteloup, 2016; Hassenforder & Ferrand, 2024).

This toolkit is a comprehensive set of participatory methods, tools and protocols designed to support stakeholders, from citizens to policy makers, to discuss, negotiate and meaningfully engage in change strategies in socio-environmental systems.

Under the leadership of IRSTEA/INRAE and in collaboration with the *Institut Agronomique néo-Calédonien* (IAC), this training was held in June 2015 and led to the creation of the first modelling in the form of a role-playing game for the Pouembout watershed (Fig. 11.2).

The first training session took place with a small group of volunteer WMC members over 2 days. This was followed by a third day of training for a larger group, including local stakeholders (farmers, customary leaders, etc.), during which the committee members who had attended the previous day's training were able to put the tools they had been trained to use into practice.

Finally, a debriefing day with the first group concluded the training. Over these 2 days of training, a range of tools from the COOPLAAGE suite was introduced, including INI-WAG, PREPAR, WAT-A-GAME, JUST-A-GRID and COOPLAN (Fig. 11.3).

11.2.2 From Training to CONCEPT Tool Development: A Shared Diagnosis

In 2016, following these training sessions, the participatory modelling tool, which was developed with Wat-a-game by training session participants, was finalised by the team of researchers and with the support of a master's student (Bouteloup, 2016). This stemmed from interest in the method and a desire to further refine the idea and improve the prototype developed in 2015. The aim was to design, with the partners, a model of how the catchment area operates and to forecast the construction and operation of the dam (Fig. 11.4).

The role-playing game called CONCEPT, was created in consultation with the WMC team. It takes the form of a board game representing the Pouembout watershed and its main geographical features, and has been calibrated based on actual

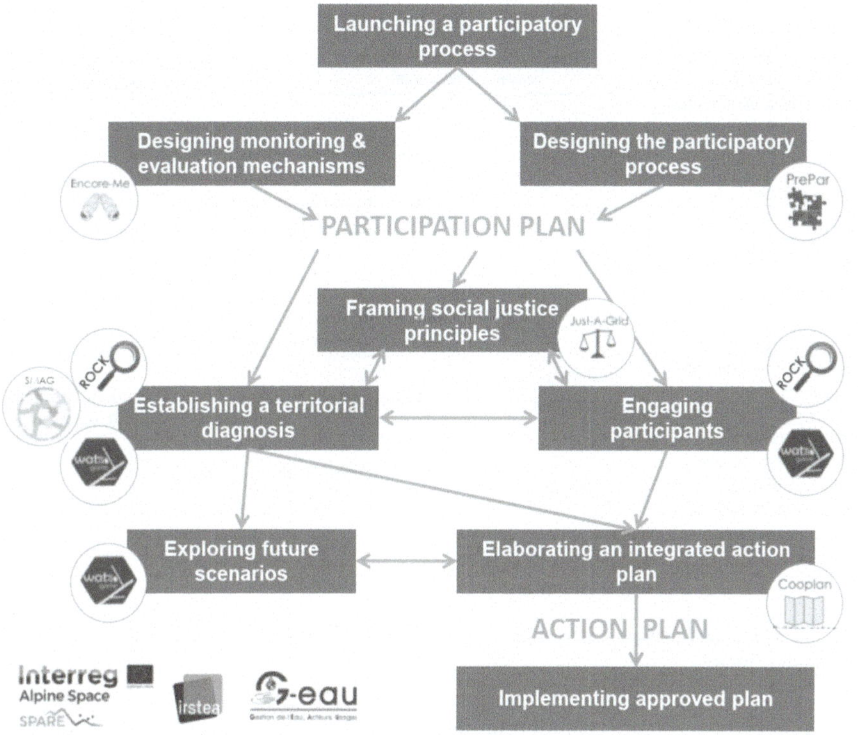

Fig. 11.2 COOPLAGE toolkit in participatory process to engage citizens. (© UMR GEAU—Irstea, 2017, http://cooplaage.watagame.info)

Fig. 11.3 Second day of training: the small group of trainees (on the left), and the day after, when the trainees present the tools to customary authorities, mayor, etc. (on the right)

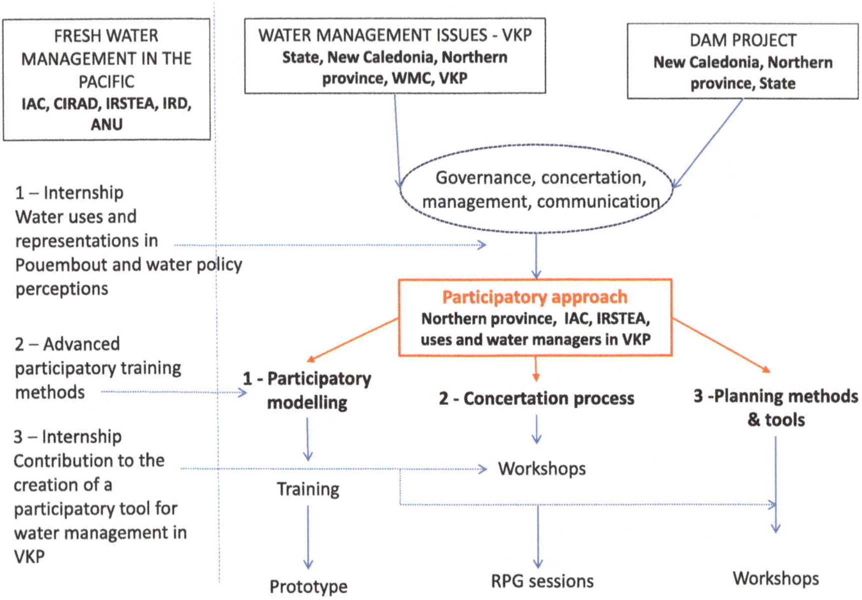

Fig. 11.4 Status of the participatory approach in the different projects. (Source: Perrine Bouteloup, 2016)

data concerning water consumption, authorised water withdrawals for agriculture and the various socio-professional categories in the area (Fig. 11.5).

The aim of this role-playing game is to share a common diagnosis in order to discuss water management. Three main objectives are to reflect on: "a global and conscientious management of water by all, future management with the dam and a broad awareness of the place and role of the manager of water networks" (Bouteloup, 2016).

The board game is made up of eight players (farmer, stockbreeder, tribal resident, village resident, entrepreneur, network manager, public institutions and public space manager), each of whom has a current activity and an innovative activity that allows the use of water to evolve over the course of the game. In addition to this, there are also four resources (clean water, dirty water, financial resources and land ownership), as well as four events that allow the game to progress (pollution, rising salt wedge, failure to maintain hydraulic structures, and failure to maintain drinking water supply networks). This role-playing game enables those present at the sessions to gain a better understanding of the various issues facing the region and to raise awareness of the challenges facing water resources. It also opens up a dialogue and facilitates deeper understanding of the mechanisms at play. Finally, setting up the game makes it possible to collect data, diagnose the situation and plan a consultation based on the data collected, as well as to explore different scenarios. The Role playing game CONCEPT was implemented and played more than 15 times by around 150 people, with distribution carried out through students from the local

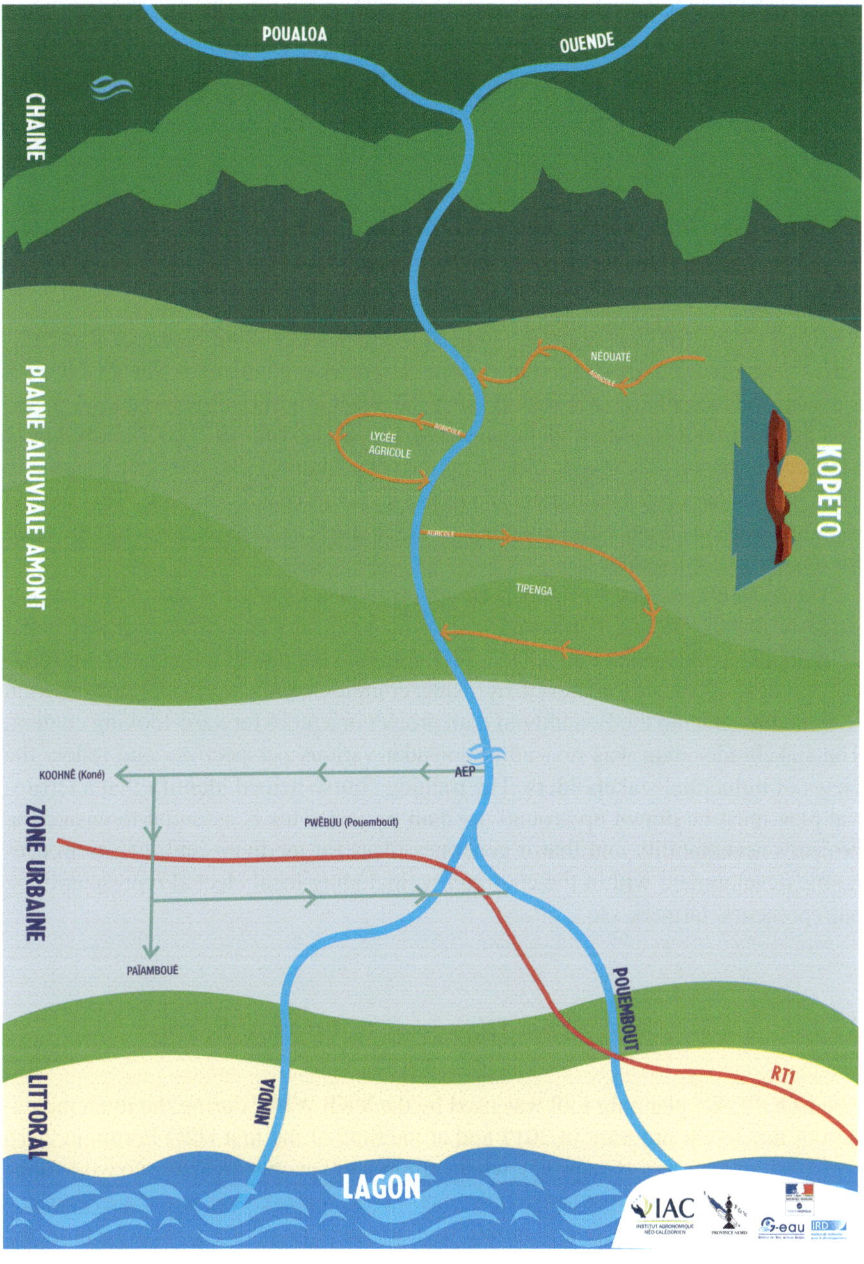

Fig. 11.5 The CONCEPT role-playing game board

Agricultural High School. This allowed the game to be tested, discussed and refined over the course of the sessions. In addition, five water ambassadors from Pouembout Agricultural High School were trained to facilitate CONCEPT. They then played the game with their families and friends.

11.2.3 The Second Round of Training: Towards Improving Planification and Facilitation

After the CONCEPT experience, the need to support co-planification of water initiatives for territorial development actions progressively emerged. The CONCEPT tool was thus supplemented within the VKP WMC by the implementation of the COOPLAN planning tool, to define the new action plan for 2017–2020. LISODE (a Montpellier-based SCOP specialist in public participation) also provided training in 2017, in Nouméa and Pouembout, on dialogue and group facilitation. This second round of sessions was complementary to the first one, clearly reinforcing the need for territorial coplanning.

These training courses provided an opportunity for the same group that participated in the COOPLAGE training course in 2017, along with several additional individuals, to incorporate the CONCEPT role-playing game into a wider consultation process. This was achieved by using complementary territorial consultation tools and by placing the Pouembout dam project in a more forward-looking context. The stakeholder map was revised to consider various perspectives and reflect the views of influential stakeholders. The training course helped identify that a territorial plan must be drawn up around the dam project. This is essential to ensure the project's acceptability and that it generates ideas for medium- and long-term economic development within the community, including local elected representatives, entrepreneurs, farmers, etc.

11.2.4 COOPLAN: From Diagnosis to Co-planification

The COOPLAN planning tool was used by the VKP WMC during the implementation of the 3-year program in 2017 and at the time of the first H2O Forum in 2018 with a wider audience (Chap. 12), where around 300 people divided into workshops used the tool.

The COOPLAN tool is a planning tool aimed at involving stakeholders in the co-construction of strategies, as stated by N. Ferrand in the MOOC TerrEau & Co. It typically involves individual or small group work to generate actions, which are then classified by the group and arranged in a planning table.

The initial stage involves using a sheet of paper divided into four sections (Fig. 11.6). The title and a brief description of the action are placed in the centre. On

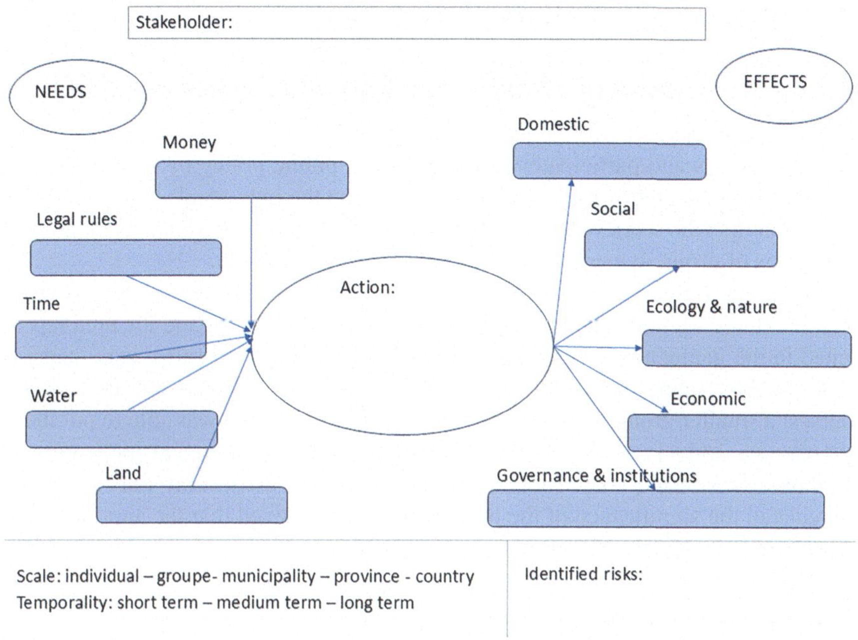

Fig. 11.6 The action sheet

the right, the impact of the result is described, including its effects on the environment, society, governance, and the economy. On the left, the necessary requirements to achieve the result are listed, including financial, human, and legal aspects. Finally, at the bottom of the sheet, the timeframe, stakeholders involved, and any potential risks are identified.

The actions resulting from this initial work are then prioritised by the entire group using stickers. The action sheets are created individually or in sub-groups and presented to explain them in greater detail.

When presenting pooled results or actions, they are displayed in a table with several columns. The table follows the general structure of the first worksheet, with action descriptions in the centre, required resources on the left, and expected impacts and effects on the right. The table may also include information on the stakeholders involved in the implementation of the action, the timeframe and an estimate of the cost of the action.

The tool was used within the VKP WMC to develop the 2017–2020 program plan. This allowed the previous years' the action program to grow and adapt by adding new themes and initiatives.

With the support of researchers, water service officers trained in the approach or involved in the VKP water management committee revitalised the COOPLAN tools as part of the development of the PEP in 2018 and 2019 (Chap. 12).

11.3 Developing a Culture of Participation

11.3.1 Assessment of Training Sessions and Supporting Tools

Training courses and participatory tools influence public policy by contributing to collaborative formulation exercises. They also affect the individuals who attend and the facilitators who work with them, on an individual and professional level, as well as in terms of group dynamics and the concept of participation.

Overall, the training courses held by VKP WMC received positive feedback from the individuals we met. The participants' perception of the tool and the area represented in the game, as well as their ability to learn about other tools, were considered valuable (interview on 1/08/19). Two groups were formed for the training courses: a smaller group that received training in the tools and was able to put them into practice, and a larger group that only participated in the practical application of the tools. As a result, some individuals in this second group may not have fully appreciated the significance of the tools utilised, or believed that the approach was insufficiently productive (interview on 03/09/19).

> While some immediately saw the usefulness of using physical objects to better understand and explain concepts that were a little difficult—sometimes it's easier when you play—some really felt like they were wasting their time playing with marbles (CGE facilitator, from the MOOC video).

Feedback on the implementation of the CONCEPT role playing game and the use of the COOPLAN in 2016, which led to the establishment of a three-year plan for 2017–2020, was also positive.

11.3.1.1 CONCEPT: Assessment

The CONCEPT role-playing game was systematically evaluated by 56 players, water users at the VKP and secondary school students studying agriculture. The tool was found to be particularly relevant (96% good or excellent), useful for understanding the problems (87%) and developing solutions (93%) (Fig. 11.7). In terms of learning, 48% said they had learned a lot about the river and its uses, almost 70% agreed that participatory modelling gave them the opportunity to think about future actions, and 46% said they had a better understanding of the practices of other stakeholders (Fig. 11.8).

The CONCEPT role-playing game, like any participatory modelling tool, reflects a complex reality. It is essential that participants are aware of the type of tool they are dealing with before any involvement in the role-play. Calibration plays a major role in the players' appropriation of the tool and in the representation of their reality. CONCEPT uses a calibration method based on key elements and figures for the area, followed by a calibration of activities through comparison to reflect the situation of the players in relation to their neighbours. This method requires a reduction in scale, which may erase details and limit subtleties, even though these elements

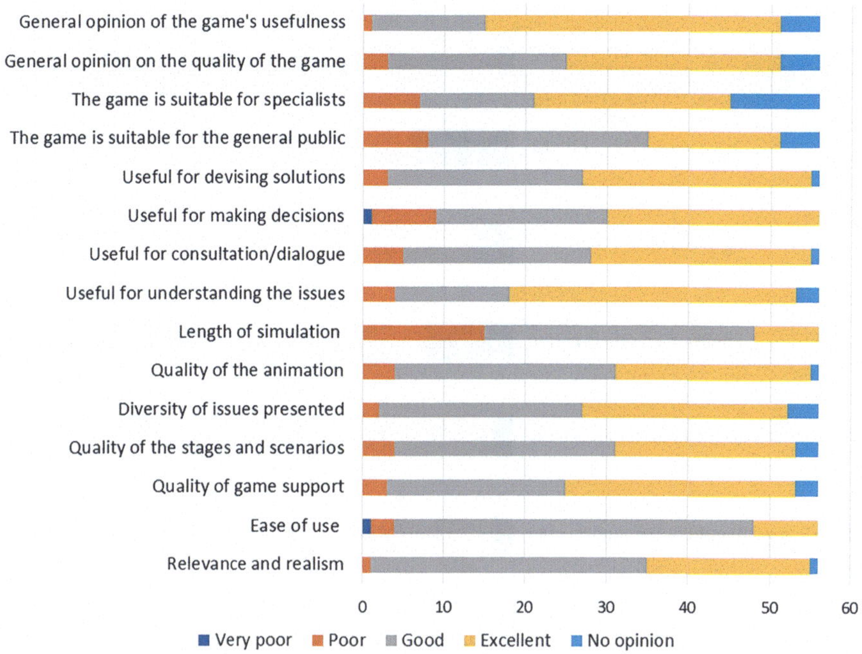

Fig. 11.7 Evaluation of the CONCEPT tool (n = 56)

may be important for the players. This may limit the support of certain players, particularly territorial specialists, who may find the model too rough. Designers and moderators need to be vigilant: some players jump to conclusions, finding that one person pollutes a lot, or that another consumes more than the average.

All the modelling and its calibration must therefore be as accurate as possible and adapted to bring out the water management themes without introducing conclusions that are too hasty or too clear-cut, which could lead to misunderstandings and a distorted representation of the current context by the players present. Above all, each modelling choice needs to be explained and justified by the designers and moderators, so that the stakeholders in the area understand the representations and validate the role-playing game as it has been constructed. It is a real challenge to strike the right balance between a game that is easy to understand and play, while at the same time being as representative of reality as possible.

During the construction and deployment of the role-playing game, a point of concern was identified. The modelling carried out so far (Hassenforder & Ferrand, 2024) has mainly focused on issues of quantity rather than quality. Although pollution is addressed through the production of dirty water and red balls in the CONCEPT role-play, environmental issues remain secondary in the model. There are several reasons for this. The participatory approach described here reflects the issues perceived by local water management stakeholders without any external viewpoints or

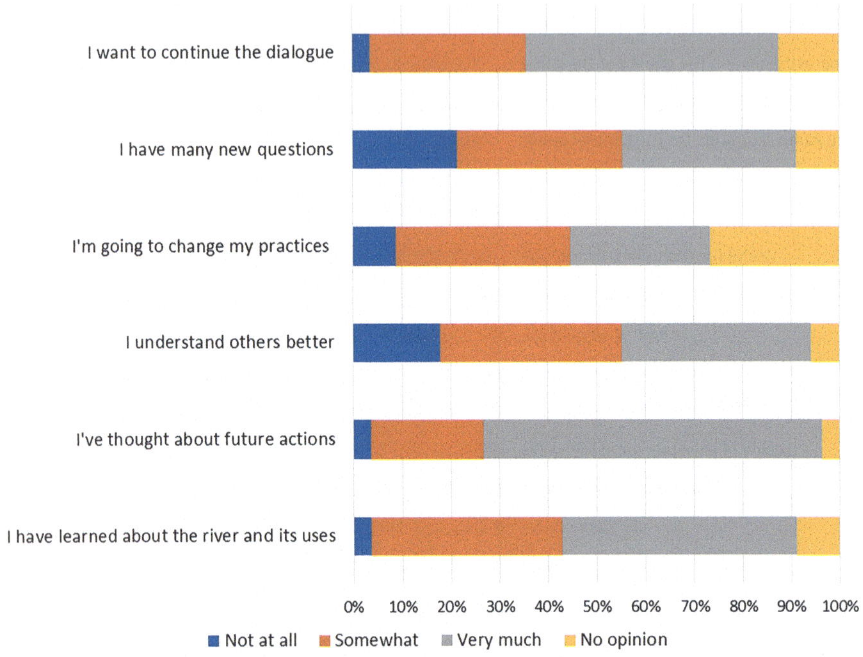

Fig. 11.8 Learning induced by the CONCEPT tool (n = 56)

opinions added. Therefore, the model hardly shows the consequences of pollution, which are a reality for various users.

The lack of prominent environmental organisations in the VKP area, in particular, and the absence of a local environmental discourse probably also explain the invisibility of environmental issues in the modelling. At the time the prototype was created in June 2015, the small and recently established environmental association Amu-keje was present. However, its dissolution in 2015 marked the end of its involvement, and the association has not been replaced. Environmental issues are addressed solely by public institutions within the scope of their responsibilities. Although the CONCEPT role-playing game inadequately illustrates environmental issues from a normative perspective, it does respect the viewpoints of local players in the area from a participatory perspective. Introducing more representation of this theme could lead to bias.

11.3.1.2 COOPLAN: Assessment

The interviewees generally regarded COOPLAN as a useful tool for planning and coordinating actions within the team. However, one participant reported receiving tasks that were not within their department's scope. The organisation of bilateral or small-group consultations with the relevant departments as a follow-up to

COOPLAN is recommended. This will help to validate the assigned actions, such as skills and workload plans.

The evaluation of the tool is limited due to the fact that the EMC's 3-year plan was not implemented as a result of a significant slowdown in the committee's activity, particularly during the COVID19 pandemic. This 'pause' limits the opportunity for a more detailed evaluation of the COOPLAN tool. However, based on the interviews conducted, it appears that the trained staff are interested in applying the tools and skills acquired in their professional careers.

11.3.2 Dissemination and Scaling-Up a Participatory Culture

11.3.2.1 Applying Serious Games Elsewhere: The Case of Maré Island

Following VKP experiment and especially the use of CONCEPT game, the customary authorities of Mare Islands, asked for increased awareness raising around water. IAC, INRAE and CIRAD thus supported the development of a role-playing game called *Tini ri rawa* (for 'water under the earth' in the Nengone language) focusing on the vulnerability of Maré's freshwater lens. Created using the Wat-a-Game participatory modelling kit, Tini ri rawa is a role-playing game calibrated for the Loyalty Islands province; it reflects local realities in terms of practices, uses and representations of water. The completed version of the game was initiated and coordinated by a student trained in participatory methodologies and tested and improved by local researchers/engineers and technicians (Amo, 2020). Based on a board representing the areas of vulnerability of Maré's freshwater duckweed, the game is based around five roles—the farmer, the craftsman, the tourist service provider, the local resident and the mayor—who must develop their activities while taking account of their impact on water resources (crop irrigation practices, rainwater harvesting, dry or conventional flush toilets, etc.).

Observations from the game sessions revealed real motivation on the part of the local population to take part in consultation and debate water management issues. By playing their roles, the players discovered other groups of local stakeholders, leading them to reconsider their relationship with water in a more global way. By putting themselves in other people's shoes, they became aware of the general interest in protecting water resources. The game session also gave each player, as an individual and as a resident, the opportunity to express their concerns, depending on their role, without any social pressures.

By transferring the game to the local authorities, it has been transformed from an exploratory scientific project into an awareness-raising tool serving the general interest (Coulange et al., 2021). The originality of this approach lies in the fact that the game was adapted to the socio-cultural context of Maré, working closely with local resource people during the development phase.

11.3.2.2 Adapting the COOPLAN Tools for the PEP

Consultation has become an essential method for planning certain projects among the trained individuals. For instance, three individuals who participated in the CGE training courses expressed their desire to reuse the participatory tools in their work, particularly the COOPLAN tool, and did actually have the opportunity to reuse it.

> One of the interviewees stated, 'I had even kept the idea of the document [COOPLAN], and I used it again to work on incivilities. You need to adapt it to the problem, but it's easy to reproduce' (interview, 20/08/19)

However, the perception of a lack of skills can be a barrier to effective implementation. The training courses were generally considered too short for the trainees to feel comfortable enough to use the tools independently.

Even so, the tools remained in people's minds, and role-playing games have become a popular method for testing governance scenarios in the context of PEP. The CONCEPT role-playing game, in particular, has been attractive for this purpose.

> Recently, the game was reconsidered due to the big questions surrounding governance and the distribution of competences in the context of PEP. Several scenarios clash, and simulation is necessary to address them (interview on 02/09/19).

In addition to learning about tool-focused methodology, the WMC's training in participation tools serves as a first step towards consultation and helps to establish a culture of consultation at both local and national levels. This was particularly important during the implementation of the PEP in 2018 and 2019 (see Chap. 12).

11.4 Conclusion

The VKP WMC's collective experience, together with research, has enabled the implementation of innovative tools for water management participation. Seeking external expertise, particularly from research, has been a key factor in the success of this initiative, leading to significant learning outcomes. The local and voluntary strategy of enlisting research to provide skills that were scarce in the area at the time has helped to build a shared culture of participation. Instead of a legal obligation to implement a wide-ranging consultation process, opting for different tools that encourage the sharing and horizontal exchange of views has encouraged managers to gradually embrace this type of approach. Currently, the VKP WMC is less active and less central to local water management. The issues of quantity have become less salient after several years of favourable rainfall. The focus has shifted towards discussions between stakeholders about sharing responsibilities and financial resources in a changing institutional framework (end of the Nouméa Agreement).

In the VKP zone, the dam project is become more concrete, potentially reducing risks to water resources in terms of quantity and focusing attention and skills on implementing the project, such as awarding contracts and advancing the territorial plan. Will the culture of participation developed in recent years be strong enough to influence the initiatives arising from these new challenges?

Even if major public investments are made to meet the region's development needs, New Caledonia's food self-sufficiency needs, and the need to adapt to climate change, participatory consultation will continue to be involved in defining public policies. Questions are already being asked, and more will emerge in the years leading up to and following the commissioning of the Pouembout dam. One such question is how to determine the best price for water. How should the rules for allocating water volumes during times of crisis be defined? What methods of agricultural development should be encouraged to balance the need to draw water for downstream drinking? These are all questions that must be addressed with local stakeholders if public policies are to be understood and accepted.

Acknowledgments For this chapter, we would like to thank Perrine Bouteloup, Léa Chevreux Warode, Delphine Coulange and Marine Pizette for their contribution to surveys, interviews and contribution to the implementation of the participatory process.

References

Amo K. (2020). *Apprendre la gestion de l'eau sur terre coutumière à partir d'un dispositif participatif : construire un dispositif pour le territoire insulaire de Maré - Mémoire de fin d'études, Master 2 Géographie.* Université Montpellier Paul Valery, 110 p.

Bouteloup, P. (2016). *La gouvernance de l'eau à partir d'une démarche participative: De l'analyse à la planification de la gestion de l'eau sur la zone VKP, Nouvelle Calédonie.* Dissertation IAC/PN, in collaboration with UMR G-eau, ISTOM.

Chevreux Warode, L. (2015). *Représentations sociales et usages de l'eau dans le bassin versant de la Pouembout.* Dissertation, Institut des Régions Chaudes, Supagro Montpellier.

Coulange, D., Bouard, S., Drouin, J., & Amo, K. (2021). À Maré en Nouvelle-Calédonie, un jeu de rôles pour préserver les ressources en eau. *The Conversation.* Retrieved from https://theconversation.com/a-mare-en-nouvelle-caledonie-un-jeu-de-roles-pour-preserver-les-ressources-en-eau-168142

Hassenforder, E., & Ferrand, N. (Eds.). (2024). *Transformative participation for socio-ecological sustainability—Around the CoOPLAGE pathways.* Quae.

Pizette, M. (2020). *L'évaluation des processus de concertation pour la construction des politiques de l'eau en Nouvelle-Calédonie.* Dissertation, Gestion Territoriale et Développement Local Université Jean Moulin Lyon 3.

Chapter 12
A Multi-level Process for Policy Design and Integrated Planning

Caroline Lejars, Séverine Bouard, and Nils Ferrand

Abstract The 'Shared Water Policy' (PEP (Politique de l'eau partagée)), it is the result of a broad participative process that involved around 1 in 600 New Caledonians. This chapter presents the multi-level participatory procedure implemented for the development of this policy, over a 2-year period. It focuses on the process, methodology and tools used to support policy design. It discusses the interests and the limitations of this process in terms of including customary values and norms.

Keywords Participation · Policy design · Co-planification

12.1 Introduction

Although water is not a very scarce resource in New Caledonia overall, its sharing and management are becoming increasingly contentious (see Chap. 3). Conflicts are beginning to emerge, around pollution, particularly from mining (Bouard et al., 2016, 2020), around over-consumption of drinking water and the payment of water bills (Davar, 2018), and around 'sacred' water on customary land (see Chap. 8). Water and associated infrastructures are the subject of growing number claims and conflicts on customary lands (Trépied, 2011; Lejars et al., 2020).

C. Lejars (✉)
CIRAD (French Agricultural Research Centre for International Development), UMR GEAU (Univ Montpellier, CIRAD, INRAE, IRD, BRGM, AgroParisTech, Institut Agro), Montpellier, France
e-mail: caroline.lejars@cirad.fr

S. Bouard
IAC (New Caledonian Institute of Agronomy), Pouembout, New Caledonia
e-mail: severine.bouard@iac.nc

N. Ferrand
INRAE, Research Institute for Agriculture, Food and Environment, UMR GEAU, Montpellier, France
e-mail: nils.ferrand@inrae.fr

C. Lejars, S. Bouard (eds.), *Water Policy in New Caledonia*, Global Issues in Water Policy 32, https://doi.org/10.1007/978-3-031-88196-1_12

Besides, the unique structure of institutions in New Caledonia has resulted in a fragmented and incomplete regulatory framework for water management that is not fully in line with local uses and practices. As outlined in Chap. 1, the management of water and aquatic environments was officially divided between four different authorities: (1) New Caledonia, which is responsible for the public river domain located outside customary lands, (2) Customary authorities, managing water resources located on customary lands. These lands, on which chiefdoms and clans exercise sovereignty, correspond to almost 30% of total land and host 52% of the territory's drinking water catchments (DAVAR, 2018) (3) Provinces, the environmental jurisdiction of which encompasses aquatic environments, (4) Municipalities (or their groupings), which are responsible for supplying the population with drinking water when water distribution is in place, as well as for wastewater treatment. However, in practice, arrangements between these different levels of administration are often favored at the local level but also undermined by current budgetary cutbacks (see Chap. 1). For instance, if the management of water resources on customary land is the responsibility of the customary authorities, the Provinces and Municipalities, after taking customary acts, can still operate on these lands.

Thus, this sharing of responsibilities, according to the type of land and coupled with regulatory shortcomings and local arrangements, makes water management at the country level very complex (see Chap. 5).

In this context, the need for establishing a strategic water framework including public stakeholder consultation has gradually emerged. In 2018, the government of New Caledonia decided to launch a broad consultation to design the 'Shared Water Policy' of the country (PEP,[1] see Chap. 10 for the detailed content of this policy). This Water Framework co-construction process involved approximately 1 in 600 New Caledonians and took place a few months before the first referendum on self-determination. The framework for the country's first water management policy was approved unanimously by the Congress in February 2019.

This chapter presents the multi-level participatory procedure implemented for the co-construction of this policy.[2] It specifically emphasizes the process and the planning support tools employed, which are partially by the COOPLAN (see foonote 2) method and VKP WMC experience (see Chap. 11). It also assesses the participatory process and analyses the challenges of including customary values and norms.

[1] Politique de l'eau partagée.

[2] Co-construction et co-planification de plan d'action pour la gestion de l'eau—Co-construction and co-planning of action plans for water management.

12.2 Main Phases of the Participatory Process Undertaken to Build the 'Shared Water Policy' (PEP)

12.2.1 Political and Technical Support for the Participatory Process

The initiative for a 'Shared Water Policy' (PEP) was launched in January 2018 and was led by two political leaders: one in charge of agriculture (non independentist) and one in charge with customary affairs (independentist). A consulting firm specialised in supporting public policy was commissioned to organise a large water forum and then to support the co-construction of the policy.

The consultants coordinated with the actors of the "MISE"[3] (Inter-Services Mission for Water), an inter-institutional group that brings together stakeholders from the three provinces: the DAVAR (Directorate of Veterinary, Food and Rural Affairs), which is in charge of the project, the DASS (Directorate of Health and Social Affairs), the DIMENC (Directorate of Industry, Mines and Energy), and representatives of the customary authorities. The MISE was established in 2017 with the objective of coordinating the technical activities of public stakeholders. However, it was only in April 2018 that it was assigned the responsibility of developing the particular policy process.

The Congress of New Caledonia officially established the political steering committee responsible for guiding, overseeing and validating MISE proposals in February 2019 (deliberation no. 395 of 20 February 2019), under the title of 'water committee'. Its structure brings together technical experts and decision-makers, including customary authorities.

The research group previously involved in participatory processes in the North Province (see Chap. 11) was appointed to support the participatory process, promoting a multi-level approach focused on social engagement, inter-sectorial coherency, process assessment and fostering conditions for implementation of the strategy.

12.2.2 A Three-Stage Multi-level Participatory Process

The participatory process and assessment of the PEP took place over a 2-year period and can be divided into three main stages: (1) A preliminary diagnosis was established and shared with different groups of stakeholders (2) the comprehensive strategic water framework was co-constructed mainly during 2018 Water Forum. This second stage was followed by (3) an assessment and multi-scale appropriation phase to include wider communities and prepare the implementation stage.

[3] Mission interservices de l'eau—Interdepartmental s mission for water.

The phase dedicated to diagnosis is covered in Chap. 10. The diagnosis was built with different groups of stakeholders (representatives of municipalities, customary authorities, miners, farmers, administrators and technicians). The sharing of data between administrations, exchanges between technicians, customary officials, associations and elected officials occurred through ten specific meetings between January and March 2018. The diagnosis was then shared with a wider audience during the 2018 water forum.

The design of the strategic framework and the actions plan mainly took place between March and October 2018. The 2018 Water Forum and the field workshops that followed were the main events that supported the policy design. In just a few months, the consultation and participatory process led to (1) the strategic water framework, a document of around 200 pages which presents the strategic objectives of the PEP and (2), an action plan with 700 actions to be implemented. The first public presentation of the strategic framework was on 17 October 2018, a fortnight before the first referendum accession to full sovereignty. It was then voted unanimously in March 2019 by the Congress.

The extension and improvement appropriation phase was conducted between June and November 2019. The previous participatory process and the water framework were assessed through a series of surveys. Taking into account the result of the assessment, field workshops were held to include wider communities in three pilot areas (especially on customary lands), to share the initial Water Framework more broadly and to re-assess locally and adapt actions so that they can be better appropriated. Results were shared at the 2019 Water forum (Fig. 12.1).

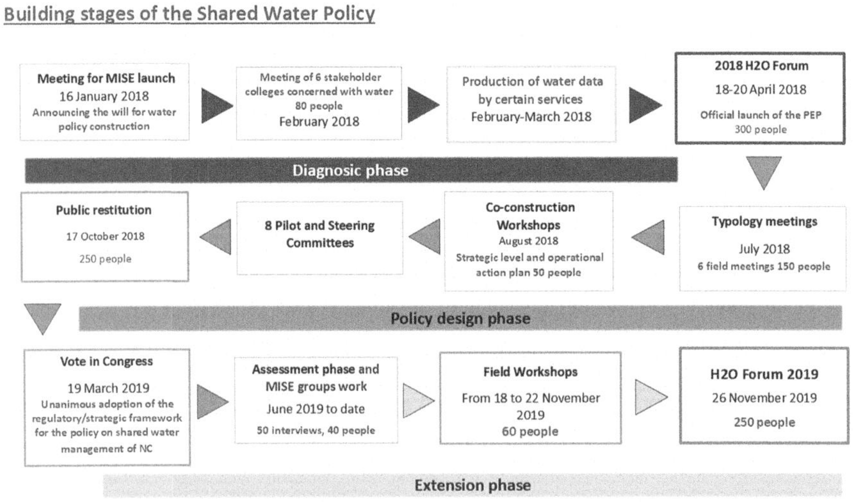

Fig. 12.1 Stages in the construction of the 'Shared Water Policy' (PEP)

12.2.3 Facing the Reality: Adapting the Participatory Process Step-by-Step

The overall co-construction process mostly followed the "classical" pathway of a participatory process: (1) production and sharing of a diagnosis of the water situation in New Caledonia, (2) proposed directions to respond to the major issues identified (3) participatory design of an action plan (4) specification of roles, governance principles and financial allocations to implement these directions and (5) evaluation of the process and the policy.

However, as in most co-design processes, the administrative officers and representatives of the communities held their own constraints linked for instance to political steering, customary rules, logistics and financial funds. As a result, the PEP participatory process was not completely linear, nor was it designed in its entirety from the outset.

Initially, in 2016, the chief officer of DAVAR was requested to structure a water policy for the country, but with a focal issue on agriculture. Then, the government required to be creative for new options, reconnect to the local authorities, and strengthen collaboration with all parties. The focus on customary lands arrived in a second time, so that the co-construction could be led both by agriculture affairs and customary affairs.

Secondly, at the beginning of the process, the objective was only to organise a Water Forum. It was only after the 2018 Water forum, and given the success of the forum, that funds to support local meetings and the appropriation phase were negotiated. The decision to continue the process was thus taken after the 2018 Water Forum, so as to include local workshops in six main regions of New Caledonia. Furthermore, it was only at the end of the first water forum that MISE was given the official mandate to continue constructing the PEP. Water Committee was also only established through a vote in February 2019, following the first public presentation of the PEP strategic framework.

Third, the research group was not included at the very beginning of the participatory process, during the preliminary consultations. The group of researchers was invited to participate in March 2018, a few weeks before the 2018 Water Forum. The consultation process with groups of representative stakeholders had already started. The specific and previous experience of the North Province with the group of researchers initiated the request for methodological support to inspire the national process (see Chap. 11).

The 2018 Water Forum was thus a key and pivotal moment in the institutionalisation of the participatory process. The extension phase and evaluation process were not planned in the first approach, essentially due to lack of budget. This extension phase was decided on and funded only 3 months after the congress vote. Therefore, once the water framework and actions plan were designed, a second round of consultation was launched, to evaluate the process and framework but also to include more citizens, adapt and continue to prioritise actions in local areas.

Throughout these stages, different stakeholders' groups were progressively included: customary authorities and citizens, farmers (a diverse community), mayors, mining industries, political leaders. It required a sound and dedicated set of methods to cope with this diversity as well as with the complexity of the water related issues.

12.3 Methodology and Tools: Adapting and Supporting Stakeholders' Involvement

12.3.1 COOPLAN-Inspired Tools and VKP WMC's Initial Experience

Despite there being no explicit reference to the forum organisers, the approach and the tools used to facilitate the forum were inspired by the Water Management Committee in Voh Kone Pouembout and the COOPLAN method (see Chap. 11). COOPLAN is part of a set of tools designed to support participatory processes and was developed as part of the COOPLAGE approach ('Coupling open and participatory tools to allow stakeholders to adapt them to water management') (Ferrand et al., 2024, Chap. 17). COOPLAGE is an integrated suite of participatory tools and protocols designed to accompany and empower groups of stakeholders towards real engagement in strategies for social and environmental change (Ferrand et al., 2024). Within COOPLAGE, COOPLAN is a method dedicated to the co-construction of action plans. Following a traditional approach, individuals or small groups first propose potential actions to implement, which are then organised and prioritised. A timetable, monitoring indicators and a provisional budget are associated with each action. The essential features of COOPLAN are its systematic elicitation of resource' use and impacts, allowing for integration and comparison of proposals, its integrative capacity for actors, scales and issues, and its accessibility for non-expert participants. The underlying principles held by the research group supporting it were: inclusion of communities, adaptation to local conditions in the country, elicitation of governance principles and transfer to local facilitators to ensure future capacities. As in most co-design processes, these principles have been discussed and shared with other groups including the administrative officers and representatives of the communities, who held their own constraints linked for instance to political steering, customary rules, logistics.

Between 2015 and 2017, some of New Caledonian water service officers were trained in this approach and these tools (Pizette, 2020; see also Chap. 11). The traditional COOPLAN method had also been implemented with the water management committee for the VKP zone (Voh Kone Pouembout, in northern New Caledonia) (see Chap. 11, Bouteloup, 2016).

During the 2018 Water Forum, the COOPLAN tools complemented the process coordinated by the consulting firm. This firm had experience in public policy construction, having notably led the design of the shared agricultural policy of the Southern Province of New Caledonia. Thus, within the scope of the PEP construction process initiated by the consultancy firm, a dedicated version of the COOPLAN method was adapted and used, with a codesigned format. But the COOPLAN process was not fully completed. This led, at this stage of the process, to a partial result, not yet addressing overall coherency.

12.3.2 *Adapting Tools for the First Water Forum*

The 2018 Water Forum played a crucial role in setting up the Water Framework (Lejars et al., 2024).

The water forum spanned 3 days with invitation extended to 800 stakeholders from the water sector. A toal of 350 people attended. One and a half days were allocated to plenary discussions and feedback from representatives of four different countries (Israel, Australia, Polynesia, France). The remaining time was devoted to participatory workshops covering four themes (Water and Environment, Water and Health, Water and Economic issues, Water and Territories). Each workshop lasted 2 h, with a participation rate of 30–80 people. On each half-day, eight 2-h workshops were conducted in parallel. The consulting firm coordinated the process with the support of a team of 24 facilitators (three people for each group, including a facilitator and a MISE representative).

Each thematic workshop followed a consistent format: a brief pre-forum diagnosis was shared through a presentation-discussion, after which a small sub-groups worked on the sheet shown in Fig. 12.2. Instead of focus on action proposals (as in the traditional COOPLAN method described in Chap. 11), the sub-groups worked on the desired outcomes (Fig. 12.2). Each sub-group was composed of three to seven people. For each proposition of "a result to be achieved", the sub-group established a timeline (short, medium or long term) and who would be responsible for its implementation (government, provinces, etc.). Each sub-group also analysed the conditions required to achieve the result, in terms of necessary skills, technical tools, regulations and funding (fitting the assessment principle of COOPLAN). The impacts on health, the economy, the environment and in terms of land use planning were considered. At the end of the workshops, each participant was invited to rank the proposals by placing a sticker on those they felt were most important for the area.

MISE members compiled a summary of all the proposals that evening, which was then presented in plenary session the following day.

Then, after the 2018 Water Forum, six additional workshops were held in six major localities across New Caledonia (called 'typological meetings' in Fig. 12.1). These additional local workshops allowed for proposals specific to geographical regions to be included and added to the proposals from the forum.

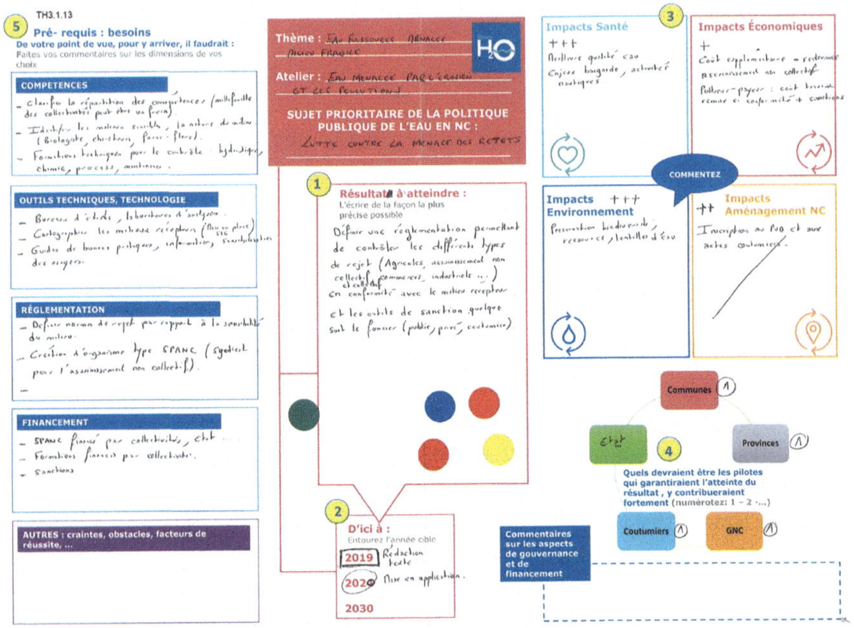

Fig. 12.2 Form used for the proposal of results to be achieved during the forum. Each group proposed (1) a result to be achieved for water policy; (2) established a timeline (short, medium or long term); (3) the impacts on health, economy, environment, and land use; (4) who would be responsible for implementation (government, provinces, etc.); (5) the prerequisites necessary, in terms of skills, technical tools, regulations and funding

12.3.3 Building the Action Plan

Based on the 2018 Water Forum and the proposed outcomes, political leaders tasked MISE actors with defining the main objectives and actions needed to achieve the results proposed by the participants during the forum and the local meetings. Internal MISE meetings were held to detail each action, identify monitoring indicators, and assess necessary budgets. Leaders and coordinators were appointed for each action, timelines and schedules were set, budgets were estimated, and priorities were assigned. MISE and DAVAR officers significantly contributed to the technical and financial details.

This effort let to detailed roadmap for implementing the PEP (Shared Water Policy, 2019). The induced shared roadmap was largely inspired by the action plan used by the VKP water management committee (see Chap. 11 and Fig. 12.3). The file and roadmap, containing 700 actions, was also used thoroughly to evaluate PEP implementation costs. It enables MISE services to monitor the effectiveness of the strategic framework implementation.

Classement OS	Objectifs stratégiques	Indicateur Impact	N°OP1	Objectif Opérationnel 1	N°OP2	Objectif Opérationnel 2	N°Action	Action	Indicateur de réalisation	Budget Investissement	Budget fonctionnement	ETP	Pilote
1	Sanctuariser nos zones de captage et nos ressources stratégiques, préserver nos milieux	Nombre de zones avec protection effective base T0 2018. Part de la population concernée par ressource protégée base T0 2018. Taux de captage / forage réellement protégés base T0 2018.	1.2	Préserver ou restaurer le bon état quantitatif et qualitatif ces eaux souterraines et superficielles notamment par une meilleure maîtrise des impacts	1.2.1	Restaurer le bon état écologique des bassins versants, des cours d'eau et des milieux aquatiques dégradés	1.2.1.8	Elaborer et mettre en œuvre un plan de remédiation prioritaire des milieux dégradés ayant un impact sur la ressource (>500 ha/an sur 5 ans)	Nbr ha réhabilités ou régulé				
1	Sanctuariser nos zones de captage et nos ressources stratégiques, préserver nos milieux	Nombre de zones avec protection effective base T0 2018. Part de la population concernée par ressource protégée base T0 2018. Taux de captage / forage réellement protégés base T0 2018.	1.2	Préserver ou restaurer le bon état quantitatif et qualitatif des eaux souterraines et superficielles notamment par une meilleure maîtrise des impacts	1.2.2	Poursuivre la protection des eaux destinées à l'alimentation humaine par la généralisation des périmètres de protection des eaux (PPE) sur tous les ouvrages AEP	1.2.2.1	Renforcer le soutien administratif, technique et financier aux communes pour la mise en place et l'actualisation des PPE	100% des PPE actualisés				
1	Sanctuariser nos zones de captage et nos ressources stratégiques, préserver nos milieux	Nombre de zones avec protection effective base T0 2018. Part de la population concernée par ressource protégée base T0 2018. Taux de captage / forage réellement protégés base T0 2018.	1.1	Connaitre nos besoins et nos ressources, définir nos ressources stratégiques : le Plan Ressources Stratégiques Eau 2023	1.1.2	Etablir un état des lieux initial (état zéro) des ressources en eau de la Nouvelle-Calédonie d'ici 2023	1.1.2.3	Définir les points de prélèvements (qualité), de mesure des débits pour l'état zéro; Réaliser des campagnes de mesures et de prélèvements nécessaires sur les masses d'eau, en priorisant sur les zones sensibles	carte des points des suivis et résultats de campagnes				
1	Sanctuariser nos zones de captage et nos ressources stratégiques, préserver nos milieux	Nombre de zones avec protection effective base T0 2018. Part de la population concernée par ressource protégée base T0 2018. Taux de captage / forage réellement protégés base T0 2018.	1.2	Préserver ou restaurer le bon état quantitatif et qualitatif des eaux souterraines et superficielles notamment par une meilleure maîtrise des impacts	1.2.1	Restaurer le bon état écologique des bassins versants, des cours d'eau et des milieux aquatiques dégradés	1.2.1.5	Restaurer les ripisylves et mettre en place les bandes enherbées en priorisant les ressources stratégiques	km de berges restauré				
1	Sanctuariser nos zones de captage et nos ressources stratégiques, préserver nos milieux	Nombre de zones avec protection effective base T0 2018. Part de la population concernée par ressource protégée base T0 2018. Taux de captage / forage réellement protégés base T0 2018.	1.2	Préserver ou restaurer le bon état quantitatif et qualitatif des eaux souterraines et superficielles notamment par une meilleure maîtrise des impacts	1.2.1	Restaurer le bon état écologique des bassins versants, des cours d'eau et des milieux aquatiques dégradés	1.2.1.9	Elaborer et mettre en place un plan de lutte contre les pollutions (diffuses, ponctuelles, accidentelles)	Nbr de projet dépollution/reduction rejet soutenu				
1	Sanctuariser nos zones de captage et nos ressources stratégiques, préserver nos milieux	Nombre de zones avec protection effective base T0 2018. Part de la population concernée par ressource protégée base T0 2018. Taux de captage / forage réellement protégés base T0 2018.	1.4	Mettre en place les outils de pilotage et de gestion de la ressource adaptés aux contextes et aux crises	1.4.2	Planifier l'aménagement et la gestion des ressources stratégiques	1.4.2.6	Définir les plans de gestion spécifiques aux ressources stratégiques, Plan RS Eau 2023	Nbre de plan de gestion RS				

Fig. 12.3 File for monitoring and prioritising actions. This Excel® roadmap includes for each main objectives of the strategic framework and for each actions proposed, some impacts indicators and budget assessment

12.3.4 Extension and Improvement Appropriation Phase: Engaging Communities and Evaluating the Process

Between June and November 2019, an extension phase was launched to assess the previous participatory process and the water framework. Field workshops were held to involve wider communities in three pilot areas (especially on customary lands in Hienghene, La Foa and Maré), to share the initial Water Framework more broadly and to enhance local relevance through additional participatory processes. A larger group of MISE facilitators from the and other social brokers were trained, governance responsibilities, including financial contributions, were discussed.

A survey was conducted to assess local knowledge and appropriation of the Shared Water Policy and to include more communities in its local design.

Three workshops in three municipalities prioritised and adapted actions from the MISE actions plan, considering local specificities. Participants proposed their own local "contributions to the PEP actions" and specified then shortly (required means, actors' roles). Then they also voted for the actions themselves (with green and red votes). They also discussed the roles of various stakeholders (country, provinces, municipalities, local groups) in decision-making process: who define priorities for the country, who decide water sharing, who contribute to analysing local water problems, etc. (Figs. 12.4 and 12.5).

Fig. 12.4 A participant in Maré votes for local proposals using stickers. Caption: "improve and support municipal action"

Case 1	Country	Provinces	Municipality	Management level	Local groups, associations
Propose and analyse solutions	D	D	P	P	

To address the objectives and challenges, the country or provinces **decide** on the types of technical, economic, or social solutions to consider. Municipalities and management structures can provide their input (P).

Case 2	Country	Provinces	Municipality	Management level	Local groups, associations
Propose and analyse solutions		P	P	D	D

OR: To address the objectives and challenges, residents and local organizations decide on the types and lists of solutions to consider. They then seek input or expertise from municipalities and provinces.

Fig. 12.5 Methodological guide showing the principle of role allocation to various categories of stakeholders (columns) for a given type of decision making, here "propose and analyse solutions". "D" marked in cells means "decide or steer", whereas "P" stands for "participate". Blank sheet means no role. Here the guide shows a comparative example (the two cases) between a solution proposal by provincial and country level actors vs. by local groups and inhabitants first

New actions were proposed and voted for in each pilot area. Many of them could be reconnected to pre-existing actions from the PEP framework and action plan. The five most voted for are listed in Table 12.1. Hienghène and Maré are areas where customary organizations are more present, whereas La Foa is an agricultural area. This is reflected in the types of actions preferred.

Participants also had to rank possible financial contributors (all taxpayers, water users, polluters) for four needs: drinking and wastewater installations, service maintenance, water provision for economic producers and catchment, rivers and environmental protection. Finally, they had to distribute ten virtual credit units among the six PEP major objectives (see Chap. 10).

This whole process was repeated three times locally plus once during the final 2019 Forum. In total, 60 people participated to the local workshops and 250 during the final forum in November 2019. These additional workshops thus support the appropriation of the Water Framework by new participants that did not take part to the first 2018 Water Forum.

Table 12.1 New actions proposed and voted by participants for each pilot areas in the extension phase

Hienghène ∑votes = 181 #voted = 52	Disseminate water quality data quickly whatever they are (10) Organize a water fiesta (10) Set field visits for children to show impacts of water and fire (10) Inform population at municipal level on water related risks (10) Renew drinking water treatment plants (9)
La Foa ∑votes = 61 #voted = 12	Strengthen actions of the La Foa Water Council (10) Stop open field irrigation during daylight (9) Support creation of water councils (7) Finding ways to promote consumers who use little water (6) Study the possibility of creating hill reservoirs for agriculture (to redirect large consumers) (6) Agricultural planning and development; assessing irrigable areas requiring inputs and defining needs and building infrastructure (6)
Maré ∑votes = 175 #voted = 55	Relaunch awareness campaigns in the tribes (12) Build one water tank per household (9) Improve the water distribution network (8) Waste management (8) Raise people's awareness (8) Inform the population about the risks (8)

12.4 Participatory Process and Water Framework Evaluation

To evaluate the participatory process, three means were used: observation, interviews and capitalization on the results. Interviews were conducted among MISE facilitators, forum and typological workshops participants, as well as in two municipalities (Hienghène and Maré) with people unable to participate in the process (Pizette, 2020). Thirty interviews were set with people involved in the development of water policy or attended to the H2O 2018 forum. These surveys assessed the participatory approach and the participatory tool used during the forum (COOPLAN), gathering opinions on its usefulness, understanding, implementation and contribution. Secondly, field surveys objectives were to evaluate the dissemination of the shared water policy beyond institutional circles and identified local priorities for action, and also to identify the population's priorities for action, in its territory, in relation to the actions defined in the water framework and in the national actions plan. In Hienghène and Maré,, interviews included mayors, directors of technical services (in charge of water) and technicians, and informal discussions, mainly at markets. Moreover, at the end of the 2019 Water forum, a questionnaire was also distributed to gather a wide-ranging evaluation. Sixty questionnaires were thereby collected.

12.4.1 Facilitators and Participants Evaluation

12.4.1.1 Feedback from Water Service Officers and MISE

Some of the MISE facilitators and staff had already received training in participation, while for others the participatory process used during the 2018 Forum was their first experience in a consultative process. Many received additional training and coaching in a second intensive round in November 2019, just before the local workshop and the 2019 Water Forum.

Many interviewees expressed pride in facilitating a multi-level participatory process, such as the 2018 and 2019 Water forums, and contributing to the design of the country's water policy. One forum facilitator noted, '*There is a sense of pride in being part of the group that participated in this success*' (Pizette, 2020).

For most water service officers, their involvement in shaping the water policy provided renewed legitimacy and gave value to their work. Facilitating this consultation process has also enhanced their understanding and knowledge of local issues through exchanges with participants. Participation allowed for open debate and dialogue, in addition to the collection of data relevant to their work. The COOPLAN tools helped facilitate and structure the dialogue. It ensured the smooth running of the workshops.

Some officers and facilitators had a more measured view. '*I have the impression that the work was done three times: during the preparatory meetings where a grid had already been proposed, then at the time of the forum where the actions were discussed again and then during the territorial meetings*' (one of the forum facilitators interviewed by Pizette, 2020). As a matter of fact, at the start of the process, the facilitators did not have an overarching view of the method to be implemented as funds were released progressively. The 2018 Water Forum was only organised with a truly participatory spirit at a late stage, the workshops on the six typological regions were only financed after the forum, and the extension and downscaling phase (three dedicated pilot areas) was decided after the vote in the Congress at the beginning of 2019. The fact that the process was set up and financed in stages created a '*feeling of repetition*'. Nevertheless, beyond the delivery of the Water Framework and action plan, the participatory process induced social learning and progressive inclusion of citizens and communities.

Participation always demands a considerable amount of time for implementation. The first step is getting organised, then engaging participants, scheduling meetings, and providing opportunities for individuals to express themselves, share their knowledges and values, and reach compromises. This process requires some back and forth, time for co-learning, as well as significant human and financial resources. In the case of the PEP, as the 'kingpin' for process construction, the MISE members took part in the water policy working group in addition to their daily workloads. Ideally, this process should have had additional human resources. '*The budget allocated to consultation in the whole construction process was too small*' (one of the forum facilitators interviewed by Pizette, 2020). A large part was dedicated to organising the forums and for communication campaigns, but only a few additional human resources were made available for the pre- and post-forum phases. As illustrated in Chap. 10, the total budget allocated to the water policy remains insufficient, even for implementing the action plan, meaning that the whole process is highly dependent on the MISE actors.

12.4.1.2 Feedback from Participants

Feedback from the Forum and field workshops was generally overwhelmingly positive.

At the end of the first H2O forum, the two political leaders in charge of PEP described it as '*a success*', '*a method that has been acclaimed*' (speech by N. Metzdorf, closing of the 2018 Water forum). Some participants viewed the process as a '*real exercise in participatory democracy*'. '*This is the first time I have taken part in a real exercise in citizen consultation and, I dare say, in participatory democracy*' (WWF Director, closing of the H2O forum).

Figure 12.6 summarises the reactions of 165 forum participants on a scale from 0 to 5: the three most agreed statements included having exchanged with people unknown before, being able to express one's point of view and that the facilitation was adapted. The three least agreed with were: feeling more engaged in water issues in New-Caledonia, better understanding of water issues and having their point of view taken into account. Several participants reported a '*feeling of sharing*'.

According to the participants, the participatory process strengthened their will to be involved in actions to face emerging water issues. For example, one proposal that was put forward by several participants was 'the "sanctuarisation" of strategic water resources'. In one workshop, this proposal was unanimously proposed by a group

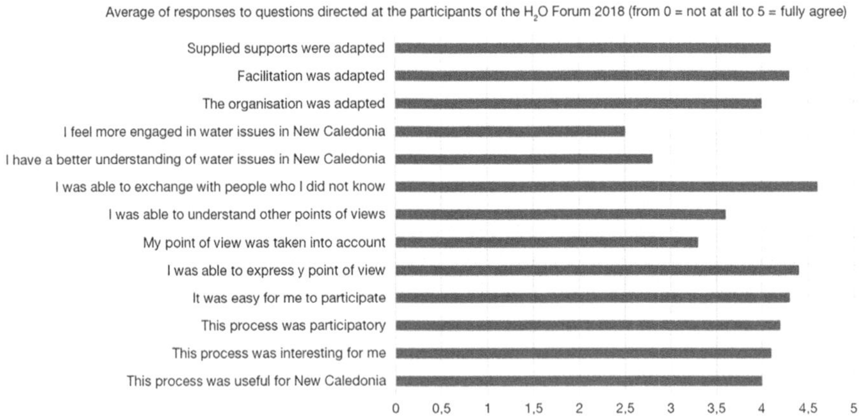

Fig. 12.6 Participants' feedback on 2018 and 2019 Water Forums

that included mining company representatives, despite the fact that such action could make mining impossible in certain nickel-rich areas. Some very innovative proposals were also put forward such as a water statute that would define water as a living legal personality, (inspired by the New Zealand's Whanganui River), and new governance systems that would prioritise the role of experts independent of political agendas.

In several groups, there was a consensus on proposals even though they could potentially create tension among different interests.

12.4.2 *Including Wider Communities and Customary Groups: Challenges and Perspectives*

During the extension phase of the policy consultation, workshops were held in three municipalities (Hienghène, La Foa and Maré) (part 12.2.4). Alongside this, local surveys were conducted with people who had not participated, especially remote tribe residents. The objectives of the PEP policy framework were shared with them and some were given the document to review and provide feedback on the resulting strategic framework.

In general, the reactions were positive: '*It's complete*', '*It's interesting*', '*Everything is said*'. The objectives of the PEP take into account the various issues encountered by the people interviewed and are consistent with their expectations. Several people noted the strong consideration given to environmental issues and the protection of resources.

However, as one interviewee put it: '*The PEP is good, but it's a bit too much of a pipe dream*' (interview in Maré by Pizette in 2020). For several interviewees, people specifically question the financial feasibility of the action plan.

The initial forum and workshops primarily involved individuals directly concerned with water management, though water impacts everyone daily. Challenges in broader participation may stem from a great diversity of factors like issue awareness or workshop remoteness. Additionally, the legitimacy to participate and the culture of participation, especially in the Kanak community, are crucial considerations.

Accessibility and distance are significant issues. For instance, at the 2019 Hienghène meeting, invitations were sent to the Haut-Coulna tribe, located an hour's drive from Hienghène, in the heart of the Central Range. Due to a bus breakdown, invitees couldn't attend. This isn't an isolated incident. Traveling to each tribe could address remoteness but might limit participant diversity and reduce opportunities for mixed debate.

Another point is relative to the participatory process itself. The participatory dimension is strongly present in Kanak communities' decision processes, in the form of "*palabres*". *Palabres* are a *discussion organised in accordance with the practices of Kanak custom and held under the authority of the clan chief, tribal chief or grand chief or, failing this, the president of the council of clan chiefs, at the request of individuals with customary civil status in order to rule on a dispute, a request for clarification or a request concerning this status or customary property. Palabres* need time, as it is necessary to '*exhaust the palabre in order to reach a decision*', and this can sometimes take several years. In such context, "*people don't interfere during discussions because they listen. It's a question of humility, but participation doesn't mean adherence, and non-participation, or absence, has no value*" (A Customary representative interviewed by Pizette, 2020). Inaction, silence, or absence during PEP meetings should not be interpreted the same way as for other cultural groups in New Caledonia.

The participatory process for the shared water policy differed significantly from traditional Kanak practices and was more time-constrained. The expectation for everyone to participate and reach a decision by the end of the day does not align with the Kanak cultural and social context. Strict protocols and structured stages in decision-making can also diminish the value of the outcomes.

In Kanak communities, traditionally, each clan and each person has a specific "role" or "responsibility" that is transmitted through lineage, language and practices. For example, some clans are responsible for water and one person in the clan can be the "master of water" (see Chap. 6), but this knowledge or role is most of often kept secret and public disclosed. This clan structure can affect the perceived 'legitimacy' of individuals or clans to speak on water issues and who can convey water-related values. Beyond the simple question of legitimacy, it also raises the question of who is able to transmit the specific values around water. It is therefore very complex to include all the representations and conceptual values of water in such a process. The legitimacy of the chosen Indigenous territories can also be questioned based on their scale or size, particularly regarding whether they adequately reflect the diverse ecological, social, and economic contexts related to water. Simms et al. (2016) highlight the difficulties raised by the scales considered for water governance in Australia (Chap. 14). The administrative (states, provinces and municipalities) or hydrographic (catchment) scales, for which institutional actors are well identified, do not necessarily coincide with Indigenous territories. In New Caledonia, as seen in Chap. 7, the representation of water is very different from the western vision. The five Kanak waters (clan water, customary water, ancient water, pure water and collective water, Chap. 7) are very different from the categories used in the PEP (groundwater, surface water, public water, water on customary lands). This raises the issue of the legitimacy of the people chosen to represent communities on these bodies.

Nevertheless, more people from Kanak communities attended the 2019 workshops than in 2018. All the customary areas of New Caledonia have been kept informed and have validated the shared water policy project, especially thanks to a person who was in charge of relations with the customary authorities within MISE. The project was presented to each of the eight customary areas and according to him, the project was '*enthusiastically*' received by the customary authorities (interview on 29/08/19). As summarised by a representative of customary authorities, '*It's important that we get involved and that the tribes also take part, that everyone takes part, especially at the moment, with the drought, it has to be everyone who pays attention to water consumption*' (Interview with a representative of customary authorities in Pizette, 2020).

12.5 Conclusion

This chapter provides an overview of the multi-level participatory approach used to develop the 'Shared Water Policy' in New Caledonia. This consultation process drew inspiration from the COOPLAN planning method and a prior experiment in VKP (Chap. 11). These tools were adapted with the support of water service officers who had been trained in the COOPLAGE approach, had applied it or had participated in a similar process on a smaller scale at a management committee level. This example illustrates, therefore, how a toolset developed by researchers can be effectively used in a large-scale consultation process.

The participatory processes described in this section have enabled innovative proposals to emerge and be discussed, particularly with regard to the status of water or the "sanctuarisation" of strategic water resources. However, whether indigenous values or representations have been more clearly taken into account and included can still be questioned, as for instance the new water law does not consider cultural and legal pluralism (see Chap. 5). Kanak representations of water were not clearly discussed during the process especially because these representations can not be easily 'operationalized' in a policy design or in an action plan. For instance, the categories of water (surface water, groundwater) used in the Water Framework do not represent the Kanak vision of water, neither in terms of categories (clan water, customary water, pure water, ancient water, collective water) as described in Chap. 7 nor in terms of 'quality' (see Chap. 9). But taking into account such categories would strongly impact water governance and would require additional knowledge to characterize each category on the ground.

Nevertheless, the participatory process has contributed to co-learning about water management, in terms of diagnosis, feasibility of action but also in terms of sharing values.

The approach allowed a consensus to be reached for the PEP, which was unanimously adopted by Congress. It was a first successful step towards consensus and sharing. But the process must continue. As Part II of this book shows, comparisons between heterogeneous communities sometimes require difficult translations and simplifications. The participative process helps to learn from each other's idioms, and from other ontologies. While the COOPLAGE tools had already been used in countries and regions with issues of indigenous claims and post-colonial conditions, including diverse communities, there are still some adaptations to be made to account better for alternative decision styles, cosmogonies and social paces.

Acknowledgments For this chapter, we would like to deeply thank Kathleena Amo, Perrine Bouteloup, Léa Chevreux Warode, Delphine Coulange and Marine Pizette for their contribution to surveys, interviews and contribution to the implementation of the participatory process.

References

Bouard, S., Lejars, C., Le Meur, P. Y., & Ferrand, N. (2016). Exploring water governance in Pouembout (New Caledonia). In *Water Governance in Oceania Workshop, Nadi, Fidji Islands, 28–29 June 2016.*

Bouard, S., Lejars, C., Sabinot, C., Coulange, D., & Grochain, S. (2020). Goutte (volet 2): La gouvernance de l'eau sur terres coutumières en Nouvelle-Calédonie: De la connaissance des usages et des savoirs à la production d'outils de gestion et de politiques adaptés. In F. Cayrol & C. Tatin (Eds.), *Au fil de l'eau, Nouméa 2019, Workshop des 17 et 18 Septembre* (pp. 30–33). PUNC, Collection CRESICA.

Bouteloup, P. (2016). La gouvernance de l'eau à partir d'une démarche participative: De l'analyse à la planification de la gestion de l'eau sur la zone VKP, Nouvelle Calédonie. In *Rapport de Stage IAC/PN, en Collaboration Avec le Cirad et Irstea (UMR Geau)*. Diplôme d'ingénieur de l'ISTOM.

Davar (Direction des Affaires Vétérinaires, Alimentaires et Rurales). (2018). *Forum H2O: Quelques chiffres clés, document présenté le 18,19 et 20 avril 2008 à Nouméa, disponible en ligne sur.* Retrieved from www.forumH2O.nc

Ferrand, N., Hassenforder, E., & Wanda, A.-G. (2024). The CoOPLAGE approach: When actors model their situation, principles or plans together for sustainable, empowering decision-making and change. In E. Hassenforder & N. Ferrand (Eds.), *Transformative participation for socio-ecological sustainability. Around the CoOPLAGE pathways* (pp. 27–41). Edition Quae. https://doi.org/10.35690/978-2-7592-3920-7

Lejars, C., Bouard, S., Sabinot, C., & Nékiraï, C. (2020). Quand « l'eau, c'est le lien »: Suivre l'évolution des réseaux d'eau pour éclairer les pratiques et les transformations sociales dans les tribus kana. *Développement Durable et Territoires, 10*(3), 22.

Lejars, C., Bouard, S., & Ferrand, N. (2024). Shared water policy in New Caledonia: Feedback on a mechanism for policy co-construction and co-plannin. In H. Emeline & F. Nils (Eds.), *Transformative participation for socio-ecological sustainability: Around the CoOPLAGE pathways* (pp. 230–240). Quae (Update Sciences et Technologies).

Pizette, M. (2020). L'évaluation des processus de concertation pour la construction des politiques de l'eau en Nouvelle-Calédonie. In *Mémoire de stage pour l'obtention du Master 2 Gestion territoriale et développement local—Cursus Monde émergent, monde en développement.* Université Jean Moulin Lyon 3.

Simms, R., Harris, L., Joe, N., & Bakker, K. (2016). Navigating the tensions in collaborative watershed governance: Water governance and indigenous communities in British Columbia, Canada. *Geoforum, 73*, 6–16. https://doi.org/10.1016/j.geoforum.2016.04.005

Trépied, B. (2011). Des conduites d'eau pour les tribus. Action municipale, colonisation et citoyenneté en Nouvelle-Calédonie. *Revue d'Histoire Moderne et Contemporaine, 58*(4), 93–120. https://doi.org/10.3917/rhmc.584.0093

Part IV
New Futures for Water Policy

Chapter 13
Water Policy in New Caledonia: Governmentality, Multi-scale Politics, and the Commons

Pierre-Yves Le Meur, Olga Peytavi, and Marie-Eve Perrin

Abstract The notion of "shared water policy" lies at the centre of the new NC water policy. Though it conveys an idea of commons, the new NC water policy is hybrid as it combines governmentality (measurement and calculation) and commons as analytical lenses offers a new perspective on emerging tensions related to water policy and the difficulties water managers and policy-makers face in crafting institutional innovations. This chapter first explores the conceptual rationale of the "shared water policy" and proposes a conceptual approach of water policy as a water-centered assemblage. Then, the focus on water policy "from below" follows the encounters of local and indigenous water conceptions and practices with new water assemblages driven by mining extraction and governmental anchoring in different localities in NC. It analyses the colonial legacy, exploring the structure, linkages, tensions and gaps between "traditional water" and "modern water" through these encounters. The case study approach ultimately highlights how local water policy entangles elements of governmentality, commodification and the commons against the background of localized fault-lines, memories and expectations.

P.-Y. Le Meur (✉)
IRD, Research Institute for Development, UMR SENS, Nouméa, New Caledonia
e-mail: pierre-yves.lemeur@ird.fr

O. Peytavi
IRD (French National Research Institute for Development)-IAC (New Caledonian Agronomy Institute), CIRAD (Agricultural Research Centre for International Development), University Paul Valéry Montpellier (UPVM), Centre de Recherche Nord-Thierry Mennesson, Pouembout, New Caledonia
e-mail: olga.peytavi@ird.fr

M.-E. Perrin
IRD (French National Research Institute for Development)-University of West Brittany (UBO), University Paul Valéry Montpellier (UPVM), Nouméa, New Caledonia
e-mail: marie-eve.perrin@ird.fr

C. Lejars, S. Bouard (eds.), *Water Policy in New Caledonia*, Global Issues in Water Policy 32, https://doi.org/10.1007/978-3-031-88196-1_13

Keywords Assemblage · Water policy · Local-level politics · Knowledge pluralism · Commodification · Governmentality · Commons

13.1 Introduction

Water has a specific position in the complex and multilayered structure overseeing public policies in New Caledonia. The governance of freshwater is shared between four institutional levels: the NC government, manager of the public river domain located outside customary lands, customary authorities, managing water resources located on customary lands, the provinces, whose environmental responsibilities include aquatic environments, the municipalities, which are responsible for supplying the population with drinking water when water distribution has been set up, as well as for wastewater treatment; the mayor is responsible for municipal policing, particularly in terms of public health (PEP, 2019: 16). These four levels do not have a clear-cut hierarchical structure. Rather they embody historical bifurcations such as the colonial spatial segregation and the land reform launched in 1978 that eventually resulted in the legal formalisation of customary land with the 1999 Organic Law and the creation of neo-customary instances (8 Customary Areas and the Customary Senate at the NC level). The process of institutionalisation of governmental and administrative instances also lasted over decades, with the creation of municipalities under French Status (1961 and 1969 laws), of the three provinces following the Matignon-Oudinot political agreements (1988), of the collegial NC government following the Noumea Agreement (1998).

The notion of "shared water policy" lies at the centre of the new NC water policy. It conveys an idea of commons—common good or common goal—which paradoxically is not extensively developed in the policy guidelines document (PEP, 2019),[1] although we find occurrences of the term and the associated notion of sharing. Actually, the new NC water policy is hybrid as it combines elements of measurement and calculation analysed under governmentality and elements referring to the commons.[2] On the one hand, beyond the notion of water as a "common good", water policy aims at building collective action and taking into account cultural specificities of indigenous water. This implies building a shared world structured around water circulation and use, an assemblage linking human and non-human actors, including water in its different guises (see Chaps. 6–9). On the other hand, the detailed description and analysis of governmentalisation procedures—discursive practices and calculating and controlling measures—transform water into quantifiable abstraction, emptied of its different meanings and substances and

[1] All the more in the context of multi-ethnic and multi-cultural common/shared destiny (*destin commun*) originating in the post-conflict/decolonization political agreements of Matignon-Oudinot (1988) and Noumea (1998) (Le Meur, 2017). The notion is absent from the "shared water policy" paper (PEP, 2019).

[2] Both concepts are defined and discussed in Sect. 13.3.

extracted from the assemblages it irrigates. The combination of governmentality and commons as analytical lenses offers a new perspective on emerging tensions related to water policy and the difficulties water managers and policy-makers face in crafting institutional innovations or dealing with changing practices.

This chapter first explores the conceptual rationale of the "shared water policy" (Sect. 13.2) by analysing how the notions of "sharing" and "commons" are used the policy guidelines document. It then proposes a conceptual approach of water policy as a water-centered assemblage that helps discuss the entanglement or intricacy of governmentality and commons in the NC water policy development (Sect. 13.3). The focus on water policy "from below" starts with local and indigenous conceptions and practices and follows the encounters with new water assemblages driven by mining extraction and governmental anchoring in Thio, located on the NC east coast in the South Province (Sect. 13.4) and beyond, in Touho and Poum in the North Province (Sect. 13.5). Drawing on local case studies, these sections analyse the colonial legacy, exploring the interlockings, linkages, tensions and gaps between "traditional water" and "modern water" that have emerged through these encounters with the hydro-social cycle—defined as "a socio-natural process by which water and society make and remake each other over space and time" (Linton & Budds, 2014: 170) or "the socially embedded techno-institutional organisation of the material flows of water" (Swyngedouw, 2015: 19).[3] The case study approach ultimately highlights how local water policy combines and entangles elements of governmentality, commodification and the commons against the background of localized fault-lines, memories and expectations.

13.2 Context: NC Shared Water Policy

In this section, we do not analyse or assess the NC water policy as it is implemented (see Chap. 12 in this volume). We rather focus on the conceptual design of the "shared water policy" (PEP) to identify discursive elements highlighting its

[3] This chapter draws on empirical data produced by Olga Peytavi for her doctoral research on the "Co-production of knowledge and institutional and material arrangements in local water governance in New Caledonia" (supervised by Séverine Bouard, IAC, Caroline Lejars, CIRAD-G-EAU and Pierre-Yves Le Meur, IRD-SENS, and funded by CIRAD and IAC), by Marie-Eve Perrin who carried out fieldwork in Poum for her master thesis in anthropology at EHESS (Budna, 2021), and by Pierre-Yves Le Meur thanks to several research projects funded by CNRT Nickel and its environment (Le Meur et al., 2021 on the value of place in mining contexts) and the Goutte Project on water governance in New Caledonia (Bouard et al., 2021; Coulange et al., 2020). Olga Peytavi carried out fieldwork in a non-mining locality, Touho (Peytavi et al., 2023), on NC north-east coast, and Thio, Pierre-Yves Le Meur has been carrying since 2008 a long-term research on the politics of resources and belongings in Thio, and specifically on water values and assemblages in 2020 with Lucie Gosset (Gosset, 2020; see also Richard et al., 2021) et in 2023 with Olga Peytavi. Two of us (OP and PYLM) carried out interviews with the staff of technical governmental services in Noumea and attended the water forum organized within the PEP process.

underlying approach. We focus particularly on the extent to which it reflects the theory and political perspective of the commons.

The PEP draft (PEP, 2019) originated in a process that started with the Assises de *l'Eau* (Water Conference) organised in 2008 by the NC Economic and Social Council (see Chap. 10). This conference already stressed the need for a strong and consistent linkages between a NC water policy in tune with reality and local initiatives under the aegis of three keywords, namely information, consultation and participation. It was followed in 2018 by the multi stakeholder H2O forum convened by the NC Government.[4] From April 18 to 20, some 250 participants worked together in workshops around four main themes: health (drinking water for everyone's health), economy (water as a source of wealth), environment (water as a threatened resource) and development (water, land, and human development (Forum H2O, 2018: 3). Finally, a draft law on NC's public water domain and protection of water resources (*loi du pays relative au domaine public de l'eau de la Nouvelle-Calédonie et à la protection de la ressource en eau*) was proposed in 2022, accompanied by a presentation report, Government of New Caledonia, 2022a, b). It was the subject of extensive consultation, as it was presented to all interested institutions (provinces, mayors' associations, consular chambers, the Customary Senate, the Economic, Social and Environmental Council) and to civil society as part of the Water Forum organised in 2022.

The discourse of the commons underlies the guiding principle of the draft law which states in its first paragraphs that "water is part of NC's common heritage" and "water use belongs to everyone". The 2019 PEP policy paper also uses the notion of common heritage (six occurrences, PEP, 2019: 31–33, 111, 155–157) and, marginally to the idea of water as a common good (PEP, 2019: 32, 122). In line with the 2008 and 2018 conferences, it promotes a common vision (PEP, 2019: 20, 94, 95), common goals (PEP, 2019: 156, 157) and common and shared orientations (PEP, 2019: 32). The notion of sharing is unsurprisingly key in the PEP given its name, but in different guises: shared vision (four occurrences), shared resource, shared diagnostic, shared management, shared consensus, and, on an ontological level, "water is sharing" (PEP, 2019: 31). Knowledge is also at stake here: "Sharing knowledge should also help build a 'common vision' of water issues" (PEP, 2019: 95). More broadly, the PEP is about "sharing knowledges, norms, and reference points" (PEP, 2019: 155). Water sharing is also associated with a notion of justice: "the fair sharing of resources (PEP, 2019: 39), "conciliation" and "fair distribution" between water uses (PEP, 2019: 75).

Sharing applies to different levels and steps of water policy design and implementation, to water use in the first place, but also to the willingness to co-construct an overall approach based on a shared diagnostic and shared orientations to develop a shared roadmap and action plan (PEP, 2019: 20, 76, 191; see also Lejars et al.,

[4] Délibération n° 395 du 20 février 2019 portant création d'un comité de pilotage pour une politique de l'eau partagée et instaurant un forum annuel de l'eau en Nouvelle-Calédonie [Deliberation no. 395 of February 20, 2019 creating a steering committee for a shared water policy and establishing an annual water forum in New Caledonia].

2021 and Chaps. 1–12).[5] The overarching goal is to "better share this source of life and wealth and ensure that it remains renewable for future generations" (PEP, 2019: 191). The 2022 draft law states in its article 1 that "Water is part of New Caledonia's common heritage, and its protection and enhancement are objectives of general interest" and "water use belongs to everyone". It also acknowledges the "Kanak and Oceanian cultural values" associated with water. (Government of New Caledonia, 2022b: 1).

The ambivalence of the term "sharing" is noteworthy. Sharing is about dividing, distributing ("sharing a cake") as much as it is about exchanging and putting together ("sharing a secret"). In brief, the PEP seeks to divide the structure of water sharing according to negotiated criteria of equity and efficiency. This process-oriented approach requires opening new spaces and/or using existing arenas for consultation and participation, but also entails control, monitoring, measurement, and assessment actions. In this way, the PEP brings together governmental procedures and elements pertaining to the discourse of the commons. What is at stake is the balance between these two orientations and the very feasibility of their combination within a consistent policy framework.

13.3 Water Policy as Water-Centered Assemblage

Water is not intrinsically a resource. It becomes a resource in the course of complex and more or less stable processes of construction, translation and assemblage. "[W]hat we call a resource or a 'natural resource' is a provisional assemblage of heterogeneous elements including material substances, technologies, discourses and practices" (Li, 2014: 589). Water policy is a particular, contextual instance of such processes, assembling actors, ideas, documents, instruments, material infrastructures and land, around "the control of the material flow of water" (Swyngedouw, 2015: 18).

How water is conceptualised plays a key role here. Water can be conceptualised as either a chemical abstraction (H_2O), a fluid and circulating substance shaping specific waterscapes or the component of complex assemblages activating a distributed agency. Water mutability renders this conceptual work trickier: "Water's diversity is, in some respects, a key to its meanings. Here is an object that is endlessly transmutable, moving readily from one shape to another: from ice to stream, from vapor to rain, from fluid to steam. It has an equally broad range of scales of existence: from droplet to ocean, trickle to flood, cup to lake" (Strang, 2005: 98). As a circulating fluid, the materiality of water is inescapable when it comes to questions of access, control and management. This materiality is also the very essence of

[5] "Create technical tools for the co-management and sharing of water for the dynamic management of water withdrawals" (PEP, 2019: 172).

water agency.[6] Water as a fluid and circulating substance that shapes specific waterscapes is seen as "the culturally meaningful, sensorily active places in which humans interact with water and with each other" (Orlove & Caton, 2010: 408). Veronica Strang crafts the notion of fluidscape (Strang, 2009) to underscore water's specific character in "how groups make use of water in material and metaphorical terms to create cultural "fluidscapes" of social connection and difference" (Strang, 2006: 149).

Water policy, therefore, is a matter of assembling, disassembling and reassembling human and nonhuman actors according to different modalities and goals. Governmental procedures conceptualise water as a chemical abstraction (Linton, 2010) in order to insert it into assemblages composed of water management services and agents, municipalities and hydrant men, biochemical laboratories, water catchments, water pipes, users, control and measurement procedures, policy documents and development projects. Water as H_2O undergoes a double process of purification (from pathogenic bio-agents and metal contaminants) and material and political embedding in water policy assemblages. We are here in the field of governmentality:

> By this word I mean three things: 1. The ensemble formed by the institutions, procedures, analyses, and reflections, the calculations and tactics that allow the exercise of this very specific albeit complex form of power, which has as its target population, as its principal form of knowledge political economy, and as its essential technical means apparatuses of security. (Foucault, 2002: 221).

Governmentality is about "the conduct of conduct" (Foucault, 1982: 220–221; Dean, 2010: 17–21 for a discussion). It is a situated way of measuring, controlling, governing populations through various instruments and apparatuses (Dean, 2010; Foucault, 2004). It is also a way of problematising a field of action—in this instance water management—as a public issue of government, inducing specific forms of subject-making (Rose, 1999; Agrawal, 2005). "Governing water means framing the problem, defining the nature of water and our relationship to it" (Fernandez, 2014: 203). In other words, if governmentality is about "the conduct of conduct," water governmentality is about "the conduct of water conduct," including who conducts the water conduct and how, to which end, and with what effects on water users as subjects.

The "assemblages" and "regimes of practices" (Dean, 2010: 40–41) defining, and generated by, water policy are heterogeneous, interspersed with different historical layers and justifications. The commons ideology is obviously an important aspect of the PEP design as we highlighted in the previous section, around the polysemic uses of the notion of sharing. The now classic theory of the commons—or rather common-pool resources—as elaborated by Ostrom (1990) is based on a set of three interacting elements: a resource, a community, and a set of rights and duties

[6] We use Wagner's view on agency here: "If the actors within a given social system are human, nonhuman, animate, and inanimate, then agency does not have to be defined exclusively in terms of intentionality, but can be defined in terms of the impact of a given actor within the network" (Wagner, 2013: 7–8).

whose interactions shape a specific form of resource governance. However, "[t]he definition of the [communities] (whatever their scale, from local communities to humankind) raises the political question of the degree of autonomy or heteronomy of a common" (Fache et al., 2021: 67). This question calls for a re-politicisation of the common, as "a contested and dynamic domain of collective existence" (Amin & Howell, 2016: 1), actively shaped through collective practices, struggles, and the crafting of new institutions (Hardt & Negri, 2009; Dardot & Laval, 2014). We can also approach the commons through ideological and discursive lenses. Wagner identifies conceptual inconsistencies in the commons literature, between common property, common pool resources, open access, or public goods, between and within academic and public discourses. He argues that "understanding commons as a social imaginary rather than as a distinct set of institutional arrangements could allow researchers to bridge these inconsistencies and thus bring greater clarity to their analytical approaches", Wagner (2012: 617).

We argue that this is the path implicitly followed by PEP. It does not take water for granted, as a mere resource to be used and distributed according to criteria of efficiency and equity. The policy document grants water "three fundamental values." Water is life, water is link and water is sharing (PEP, 2019: 32). These statements clearly diverge from simple utilitarianism and even discussions of justice, and express an ontological positioning. However, this "commons imaginary" (Wagner, 2012) is not systematically translated into policy measures in the rest of the text though it paves the way for recognition of "Kanak and Oceanian cultural values'" associated with water as stated in the first article of the 2022 draft law on water (Government of New Caledonia, 2022b: 1).

13.4 Tinkering Water Assemblages in Thio

To document the intricacies between governmental practices and commons imaginary in water policy, we will explore how water assemblages are tinkered (Peytavi et al., 2023), reshaped and enacted in different settings in New Caledonia. We first focus on Thio and analyse the progressive implementation of water payment in this historical nickel mining stronghold in New Caledonia (Le Meur, 2024) against the backdrop of multi-residential and segmented approaches.

13.4.1 What Paying Water Means: Governmentality, Commodification and the Production of a Municipal Commons

> Mentalities are changing, but the idea that 'the water belongs to us' is still there. However, now the water is indeed there, but it's the tap water and the water pipes that make the difference. Tap water and pipes come from the white colonial settlers, and when you go to them,

> you pay. So, there was a debate, because here they didn't want to pay for the water, as it was the water from Saint-Pierre, from Saille Forest. Okay, the water comes out of the tribe,[7] it belongs to the tribe, but the reservoir and the water pipes, the tap for it to reach the house, that's something else. Otherwise, we go back to how the old folks did it, we'll cut bamboo and make our own pipes, we'll fetch water with buckets. But now, who's going to fetch water with a bucket? That's over, those times are gone. We all need to have toilets in the house, a sink, water, a shower. Even showering outside like that, it's over. (Interview with a municipal officer, Thio, 2020)

This interview excerpt brings together in a very evocative way water history, governance and property and the complexity of their intricacies. It also conveys a sense of the assemblages that water is part of and how water policy is embedded in, and reshaped by local-level politics and social trajectories. As we will see, segmented approaches shaping Kanak politics (Demmer, 2009), multi-residential strategies (Le Meur, 2013) and equity issues in land development and infrastructure distribution at the municipal level (Bouard, 2013; Herrenschmidt & Le Meur, 2016) influence water policy design and implementation at the local—meaning municipal and infra-municipal—level. Furthermore, material and monetary issues are at the centre of water supply and access in Thio as in other NC localities (Trépied, 2011; Peytavi et al., 2023).

The history of how water payment was implemented in Thio and what reactions this innovation triggered among the population forms a useful entry-point to local water policy. The meaning of water payment has sparked heated debates, alongside the basic question of acceptance which has also been widely discussed.

"*Who owns water should not pay for it*", here is the basic and often heard argument implying a strong sense of local sovereignty. In this respect, water payment is seen as a mark of dispossession. The issue is related to the uneven spatial integration of the territory in Thio: the tribes of the Borendy district (coastal area located 30–60 km south-east to the village), of Koua, Kouaré, Ouindo (in the central hills) and Ouroué (north-west to Thio on the coast) feel overlooked in preference to the village and the nearby tribes of Saint-Pierre, Saint-Paul, Saint Philippo I et II who are all connected to the new Saille water catchment. The residents of these remote tribes do not understand why they should have to pay when the water supply is so rundown. This sentiment is reinforced by differences in water management among tribes and even within the same tribe. The notion of a municipal-level water commons has no appeal for them and they tend to value people having their own water spring and taking care of it, or tribes managing their own localised water supply. The Saint-Joseph tribe in Borendy manages its own water source outside any municipal supervision and the inhabitants do not pay for water. Owning the resource and not having to pay are intrinsically linked and together constitute a claim for local sovereignty on the issue and a rejection of the interdependence implied by shared water management at the municipal level.

---

[7] The Neo-Caledonian notion of tribe (*tribu*) refers to a historically produced administrative unit created in 1867 by colonisation and now used in everyday language as the equivalent of Kanak village.

However, paying for water access and use can be seen the other way round, as the sign of, and a means for regaining control over water as a local resource managed at the municipal level, thereby being a question of local sovereignty. Water commodification is not just a matter of money if the implementation of shared water system payment is aimed at funding the local capacity to manage this resource in a sustainable and equitable way by facilitating water network maintenance and ensuring the self-sufficiency of the community.

To fully understand these debates, we need to understand the trajectory and transformations of Thio waterscapes and examine historical and geographical factors including territorial continuities and discontinuities, as well as the spatial impact of water policy. The violent civil clashes of the 1980s—euphemistically called "the events"—constitute a pivotal episode in local memories and discourses, marked by a before-and-after and reminded as crystallizing a turning-point or new beginning, all the more as Thio was a hot spot of this sequence in New Caledonia.[8]

The stories are still very much alive for those who lived through this period of insurgency, including bad memories of the mining companies' practices prior to "the events" (Levacher & Le Meur, 2022; Peytavi, 2024). The fact that the tribes started to get a better access to water precisely at that time also results from the inhabitants' active participation: they actually seized machines on the mining sites to build infrastructures, mainly roads, and engaged in negotiations for water access.

> You see that road there, the one you took to come here? There used to be no road; the road used to go up the mountain over there, far behind. In 1984, we had to fight to get shovels, trucks, and all the equipment from the SLN.[9] It was us who built the road there all the way to Petit Borendy. When we occupied Thio, it was us who went to get all the equipment, taking it from here and there. (Interview with J., Tribe of Port-Bouquet, 2022)

The 1980s—precisely "after the events of 1984"—are seen as marking a significant change with the start of an explicit "water policy" defined by specific social and health parameters (equity in water access, chlorination). However, for Botamere, Pétroglyphes, Nakalé (SLN mining villages), running water was available in homes by the 1960s–1970s. Before the 1980s, only the village and the mining compounds had access to drinking water and the idea of a new water policy actually meant the takeover of the issue by the municipality and the extension of infrastructure supply toward remote (and not so remote!) tribes. SLN's networks were gradually ceded to the municipality as the institutionalisation of this body took shape, notably with the Billotte law of 1969.[10]

---

[8] A pivotal event as a moment of reconfiguration of the relations between past, present and future (or regime of historicity, Hartog, 2003) is also a diagnostic event (Moore, 1987) for what it reveals of these transformations.

[9] Société le Nickel, a historical mining company in NC and nowadays a subsidiary of the French mining group Eramet.

[10] Loi n° 69–5 du 3 janvier 1969 relative à la création et à l'organisation des communes dans le territoire de la Nouvelle-Calédonie [Law no. 69–5 of January 3, 1969 on the creation and organization of communes in the territory of New Caledonia].

Subsequently, each tribe collaborated with the municipality to build its own water catchment, with the active participation of the inhabitants. This collective effort resulted in the creation of over 15 water catchments serving the tribes and mining villages. During this period, the idea of building a single water catchment to serve 5–6 tribes began to take shape, targeting the Saille forest water catchment. By the 1980s, the municipality commissioned studies on the water resource in Thio (including the Saille catchment), under the aegis of Louis Mapéri, the first pro-independence mayor of Thio. (He was elected mayor in 1986 after 1 year of direct rule by the French government in the aftermath of "the events.") The Saille project was born but it would take a long time to achieve its full implementation. The project effectively started under Thierry Song's municipal mandate in 2013 and was completed after Jean-Patrick Toura was elected mayor in 2014. The water catchment was finalised in 2016 and it now links the tribes of Saint-Michel, Saint-Paul, Saint-Pierre, Thio Mission, the SLN villages of Pétroglyphes and Nakalé, and Thio central village to the same water supply. Saille replaced a total of five water catchments.

> Now that we have gravity-fed water in the Saille water catchment, it's less expensive for the municipality. We've eliminated many sources to keep only one source. This will significantly reduce costs in terms of maintenance. Because, just in Thio, if I count St. Paul, St. Michel, we have around thirteen tribes, approximately 20 water sources. That's too many. If we have one source that can supply 4, 5, 6 tribes, a neighbourhood, and the village, financially, it's good for the municipality. That's no longer just the tribe's water; you're entering a different dimension. That's why we say the water comes out of the tribe, but the water pipes, the service, it goes beyond that now. All of this is a matter of financing. (Interview with one of the Thio hydrant men, municipal officer, 2020)

The hydrant man bears a discourse of technical rationality and economic efficiency, in short of governmentality. He also suggests a change in water assemblages, including pipes and services, that results from this process of governmentalisation of water management: "*you're entering a new dimension.*" This justifies the introduction of a payment system to maintain the pipelines. The communal water officers provide a rationale for the water charges by emphasising their regular monitoring of water quality and quantity. They conduct monthly water sampling and testing, utilising the services of a private laboratory, which incurs expenses. The resulting data is accessible to the public at the municipal office.

The meticulous participant observation of municipal hydrant men's daily work by one of the authors (Olga Peytavi) provides more specific insights into how these processes take place on the ground. She followed the two municipal hydrant men in their different tasks,[11] conducting repairs, cleaning water catchments, taking water

[11] The two municipal hydrant men are Kanak inhabitants of Thio. Both belong to the Koua tribe but reside in different areas, one in Thio-Mission and the other in Nakalé. The senior member has been employed as a hydrant maintenance worker since the 1987, offering invaluable experience regarding the evolution of the water system. He assumes the leadership role among the hydrant maintenance staff, dictating daily operational tasks. Specifically, their duties include cleaning the water catchment and inspecting facilities on Mondays and Fridays, while the rest of the week is spent on standby, responding to residents' reports of leaks and promptly repairing them. At the end of each

samples to check for quality. Even though the water catchment is far away, municipal officers find it is easier and less time consuming to clean one recent and modern water catchment than the five old ones. Hence the project to link the isolated tribes of Ouindo and Kouare to this water catchment. There is a second project in the Borendy district, aimed at linking the Grand Borendy water catchment to most of the Borendy district tribes. However, this is not without local resistance (see below). In the long term, municipal officers would like to have only two large water catchments supplying the whole municipality, maintaining the old water catchments as emergency back-ups.

The discourse of municipal officers about rationalisation goes hand in hand with a political objective of achieving water self-sufficiency and sovereignty. For them, having a municipal water system allows the inhabitants to be self-sufficient, independent from other localities. They can manage water collectively and achieve municipal sovereignty. Water as a municipal commons and tool for sovereignty is produced through water management rationalisation and commodification. The introduction of water payment is not intended to generate revenue from inhabitants, but at financing the infrastructure, staff, services and technical equipment required to maintain general access to these water pipes.

For example, on St. Benoit Islet, one of the village neighborhoods, which is actually a SLN slag heap and home to many inhabitants from the Borendy District, the local administration abstains from charging for water usage due to the antiquated nature of the existing pipelines. Consequently, there are plans underway to overhaul the pipeline system, similar to the broader initiative to interlink Saille and Kouaré or the project of a unified catchment system at Borendy. This scenario highlights the municipality's viewpoint, wherein the provision of water services is regarded not as a pursuit for financial gain, but rather as an intrinsic communal obligation. However, while water service providers advocate strongly for regular payments, there are numerous exceptions throughout the municipality, leading to a degree of ambiguity in this stance. These exceptions are frequently attributed to inadequacies in the infrastructure, highlighting the complexity of balancing service provision and financial sustainability within the context of municipal water management.

In this respect, water payment is a shared water payment to collectively manage water and water access. This is reinforced by the communal decision to build collective water catchments providing for two to five tribes each.

This reflects the lack of strict opposition between governmentality, sovereignty and the logic of the commons; instead there are intricacies, feedback and co-construction. It is worth noting that they do not reject water's cultural value or replace it by monetary value; instead they add a monetary layer on the cultural layer, implicitly asserting—it might be over-interpreting—that the preservation of water's cultural value requires this process of economic rationalisation. In this regard, culture is conceived as the building block of sovereignty.

---

month, they conduct quality assessments on each water catchment to ensure water purity. Additionally, every 3 months, they inspect water meters to facilitate accurate billing for consumers.

> The history of the sources, their stories within the tribes and clans, remain because it's the history of each clan, each tribe in Thio. But the use of water has changed, from sources, it has changed to tap water. Today, it has not only cultural value but also financial value. This dimension must be considered because as we move forward in time, the financial value of water allows each municipality to manage itself. Water is one of a municipality's assets. It's like nickel. We use our own natural resources to finance our own municipality. We self-sustain with our water. (Interview with one of the Thio hydrant men, municipal officer, 2020)

However, it's crucial to note how water intertwines with trajectories of the Kanak clans within Thio's historical narratives and collective memories, including the emphasis placed by municipal officers on their significance and role for Thio's inhabitants. Nevertheless, they draw a line between this water and the tap water used for drinking and household tasks. These two types of water hold distinct attributes. It's not a matter of one having more or less value than the other, but rather, they are ontologically different, utilised in contrasting contexts and for different purposes. The commodification of tap water is one of these contrasting features and contexts, whereby money becomes a component of the municipal water assemblage.

13.4.2 Local Water Politics, Segmented Structures and Multi-residential Patterns

Local water policy combines and entangles elements of governmentality, commodification and the commons. It also reveals different modes of existence or ontologies for water, depending on the type of assemblage it is part of. Other factors contribute to making this waterscape even more complex, namely social and political segmentation logics and multi-residential patterns.

The differentiated methods and rhythms of incorporating the territories that make up Thio into the current municipal water policy has generated tensions in different places and at various scales. This leads to paradoxical discourses and pluralistic practices. On the one hand, many people criticise the municipality for its perceived lack of investment in the water supply and sanitation system and the fact that some tribes and areas are excluded from its modernisation. On the other hand, people promote self-sufficiency and independence from the municipal water system. We will outline a few localised cases to substantiate this argument.

13.4.2.1 Koua and Thio Mission Tribe: Historical Multi-residential Patterns

When asked, the municipal water officers told me (Olga Peytavi) that they pay around 6000–8000 XPF (around 50–70 Euros) for personal water use per trimester. The family I am staying with in the Port Bouquet tribe gave me a similar figure: around 6000 XPF (50 Euros) for one trimester. However the municipal officers both have a house in the tribe of Koua, their birthplace, where none of them is connected to the municipal water network. They take care of their own source of water and are proud of it. Actually, the majority of Koua residents living downhill are not connected to the municipal water system. Instead, they have established their own water supply system, obtaining water from a nearby small creek.

The Koua catchment is an aging water catchment constructed back in the 1980s. However, in the early 2000s, a cyclone caused significant damage to the reservoir. Since that incident, the municipality has refrained from billing Koua's residents for their water consumption. There are plans to build a new water catchment and reservoir, but the precise timetable for this project remains undetermined.

Koua has a special history. The area was emptied of its inhabitants after the 1878 rebellion led by Ataï. Some of them fled in neighboring localities, others were deported to the Isle of Pines south to NC main island, and came back more than a decade later.[12] Thio Mission was created around the main church in the 1870s gathering Kanak people from different localities in Thio (similar to a missionary settlement) became the main residential hub for people coming from Koua. With the land reform launched in 1978 and intensified in the 1990s, land was distributed in Koua to former clan members in the form of customary land. Nowadays, a significant portion of its inhabitants do not permanently reside in Koua. They live in the Thio Mission tribe, also known as Saint-Philippo II, next to the village. These residents return to Koua on weekends or during holidays where they have their yam fields and feel at home. This of course also generates difficulties for the municipality regarding the material and financial management of infrastructure supply and maintenance, including water and other networks.

13.4.2.2 Borendy District: Negotiating Remoteness and Water Self-Sufficiency

We find a similar multi-residential pattern for the tribe of Petit-Borendy and Saint Jean-Baptiste, the most remote tribes of Borendy district, about an hour and a half away from the village. The inhabitants are partly connected to the communal water catchment, while others have their own water source. Most inhabitants of these two tribes residing in the Thio Mission live in what they call "*baraque*" (shack), which is a temporary, precarious form of housing built from metal sheets. Even if people live there most of the year, farm their yam and taro fields there, the place they actually relate to is the tribe they are from. This move from the faraway tribes is reinforced by the deficit of infrastructure (hence the surname of the area: the "forgotten coast"): bad roads, no communal school, lack of water system.

The Saint-Joseph tribe, also located in the Borendy district, comprises approximately ten families, meaning around 40–50 residents living there. However most of them work in Thio village or Nouméa. Only about ten residents live in the St-Joseph tribe full-time. The St-Joseph tribe consists exclusively of the Oundo clan. The

[12] This was more generally the case for the clans living in the interior of Thio, in Kouare and Kouergoa in the neighboring municipality of Bouloupari (Le Meur, 2013).

Oundo people arrived from Ouenghi, on the other side of the main island, with some of them stopping at Thio-Mission, while the rest continued to St-Joseph. Their water supply comes from a spring that was established by the elders in the 1970s, as recalled by the residents. They do not have a reservoir, and the houses are directly connected to the spring's water catchment. Therefore, there is no chlorination. It is "pure" water, the "natural" water from the spring, as the interviewees emphasise. When the spring gets clogged with leaves, young people from the tribe (both girls and boys) climb up to unclog it. The residents have not encountered any major problems, except for water levels dropping during the dry season. In such cases, the elders convene and advise residents to reduce their water consumption. They do not have wet taro fields and practice rain-fed agriculture. Only the flowers are watered using the pipe water. This is an obvious example of an infra-municipal segmentation strategy to support self-sufficiency and adapt to the remote location of Saint-Joseph which is (so far) outside the reach of municipal networks.

Actually, in the 2010s, the tribe inhabitants refused to be connected to the Grand Borendy water supply because they wanted to remain independent. They argue that since the municipality does not maintain their spring, they do not pay for water. However, the pipes are starting to get old, and there are still old cast iron pipes at the head of the network. These pipes have not been renovated since the network was installed in the 1970s. There are many leaks, and the residents of the St-Joseph tribe are forced to repair them with rubber bands. They asked the municipality for help in renovating their spring, but their request was refused as a sort of retaliation: the municipality refused to assist them for they had refused to be connected to Grand Borendy. The tribe was willing to cover the repair costs, but the project fell through and was never carried out.

However, the municipal water officers still take water samples from the communal house tap to check the quality. According to the water officer, there are no quality problems to report. For residents of other Borendy tribes, St-Joseph appears as a success story of management based on the least possible involvement from the municipality, an example of local sovereignty in water matters.

13.4.2.3 Ouindo and Saint-Paul Tribes: Navigating "Modern" and "Traditional" Water

Ouindo belongs to the group of tribes located inland, in hilly areas away from the main road and Thio central village. It illustrates the contrast between the water system expanding around Thio Village, especially with the implementation of Saille water catchment, and the adaptive strategies developed in such outlying areas.

While accompanying the water officers during her fieldwork, Olga Peytavi heard them consistently emphasising the significance of water payments and highlighting the collective responsibility of the residents. However, the Ouindo water catchment visit revealed that a single water meter serves the entire tribe. The question of who is responsible for payments and whether anyone is actually making these payments remains unclear, and this does not appear to be a matter of concern for local people nor for municipal agents. The size of the Ouindo tribe might explain this lack of concern. Only 50 people live in Ouindo, in contrast to the 592 people residing in the Thio-Mission/Philippo II tribe (ISEE, 2019).

One of us (P.-Y. Le Meur) also observed in another tribe of the central valley (Saint-Paul) in 2020 how people were forced to literally tinker their access to water due to the disruption in traditional water circulation induced by mining-related environmental damage from the 1950s and 1960s when extractive capacities brutally increased with the mechanisation and motorisation of nickel mining (Gosset, 2020: 19–22). Traditional assemblages linking humans and nonhumans, taro irrigated fields, bamboo pipes and domestic water, and also creek shrimps, eels, and ancestors, were replaced as such but some of their components remained to be plugged into new tinkerings striving at connecting municipal, mining and ad hoc pipes in innovative ways. Old water springs were kept active as they remained embedded in the history of their foundation and specific practices and tabu, while simultaneously being linked to "modern" water supply systems, be they municipal or, before that, in the 1960s, implemented by SLN, the main mining company in Thio. The damage caused by mining and the resulting impossibility of using traditional open-cast pipes led the Saint-Paul chief to negotiate compensation for the tribe with the head of SLN in Thio. The company installed a pipe from the spring (at one of the mother spring's resurgences) across the creek at ground level. This pipe stopped at the beginning of the open-cut conduit. Residents continued to take water from the open pipe. During floods, however, the tribe's water supply was disrupted: the pipe in the creek broke, and "pollution" (red soil from the mine) was carried into the pipe and into the taro fields. The mining company was responsible for repairing the pipe in the creek where people from Saint-Paul were responsible for cleaning up the pipe: "Because we take the nickel out of the pipe with the shovel. We make an opening to let the mud out" (man, Saint-Paul, 2020). Before the municipal water distribution was progressively put in place from the late 1960s and more definitively in the 1980–1990s, Saint-Paul inhabitants had to alternate between using water from the pipe when it was operational and taking water from the river during periods of rain and cyclones.

> Sometimes when the weather's bad, the canal's cut off. We can't use it anymore because it's clogged up with pollution. The water can't circulate anymore, so we have to fetch water from the river. The elders go down there and try to clean it up so that the water still comes to the tribe. (Man from Saint-Paul, 2020)

13.5 The Political Entanglements of Water Policy in Thio and Beyond

This section aims at completing the focus on Thio through brief insights in two other localities in New Caledonia, both located in the North Province: Touho, a non-mining site where one of us (Olga Peytavi) carries out her doctoral fieldwork, and Poum, with a double issue of containing water on the mining site and bringing it to the inhabited islets, where one of us (Marie-Eve Perrin) did her master thesis.

13.5.1 Thio: Discontinuities and Changing Assemblages

The way water access, supply and maintenance are organised in Thio underscores spatial discontinuities and inequalities. It also reveals individual and collective agencies at different scales. There are strong discrepancies between Thio village and nearby tribes on the one hand and remote areas located in the central chain (Kouare, Koua, Ouindo tribes) and to the south-east on the coast (Borendy district). Another discontinuity concerns mining sites where water is managed autonomously and according to specific criteria.

This contrast underscores the tensions related to governance, encompassing issues such as payment, quality and quantity control, and the spatial integration of different tribes based on their proximity to the village. Residents express skepticism and lack of trust in the municipal authorities. However, while they criticise the municipality for their perceived lack of commitment, they commend fellow residents or tribes who have achieved independence regarding water access or employ alternative water sources.

Water flows highlight discontinuities and disruptions in the landscape partly resulting from spatial inequalities in land development at the expense of the most remote areas. As we have seen, there is also a segmentary approach at work, partly as adaptive responses to perceived municipal shortcomings. As one of us (Olga Peytavi) remarked in her field notebook, "one of the first things that caught my attention upon arriving in Thio was the abundance of privately owned water catchments scattered throughout the area, each serving one or two families." We encountered many accounts from residents who alternate between the new Saille water catchment and their tribe's older water catchment according to a risk-avoidance strategy (relying on multiple water sources) and in an effort to lower their costs.

This risk-avoidance strategy also stems from the unpredictable but recurrent damages caused by nickel extraction, especially from the 1960s onward with the increase of mechanisation and extractive production forces. Several accounts mention the surprise and shock of seeing mining pollution in the canal. The difficulty of cleaning the canal with the red earth was also mentioned: "The first pollution of the canal was so strong that it stayed in the canal."

Nickel extraction is not only the material cause of the disruption of traditional water distribution. It has also contributed to shaping distinct imaginations and expectations, representing the psychological aspect of mining landscapes (Ey & Sherval, 2016) and the ideological layer of the mining enclave (Banks & Le Meur, 2024). Thio has been a historical stronghold of nickel extraction since the 1870s (Bencivengo, 2010; Le Meur, 2024). People establish a sort of parallel between nickel and water emphasising the importance of natural resources and the need to regain sovereign control over them. Nickel even functions as an ambiguous "universal measurement unit" of development in Thio. A recurrent discourse is that, while nickel has brought jobs and income to people, it has not generated development, meaning Thio has not benefited from nickel revenues at the collective level as shown by the poor level of infrastructure development. A functional water distribution system is seen as a sign of development and, conversely, its uneven presence and the neglect of remote areas as an indication of failed development.

In the meantime, other assemblages around water arise from the convergence of nickel extraction, bushfire impacts and the erosive effects of invasive species (deer, pigs) with the geological features of subsoil, in the form of chromium 6 contamination observed in diverse locations (Borendy, Saint-Paul for instance), making the option of using old catchments trickier (Even with the new Saille catchment, Saint-Paul inhabitants occasionally stayed without water for several days.) Recent analyses have shown a chromium 6 content higher than the authorised standard for water to be used as drinking water and this raises scientific issues about impact on human health and the role of nickel extraction in this problem (see Chap. 4). This indicates the difficulty in purifying water as H2O and containing its circulation and participation in various and changing assemblages, a key issue in Poum too (see below).

13.5.2 Touho: Paying or Not Paying

In Touho, a rural commune situated in the non-mining eastern-northern region of the country, water payment operates on a fixed-rate basis, set at 5000 XPF (around 40 Euros) per semester. The payment of water bills is more or less carried out by the residents. The municipal administration aims to transition towards a consumption-based payment model, as evidenced by the installation of water meters in each household by water service personnel, although these meters are not yet operational. The presence of these meters, however, introduces complications, as they frequently become clogged, necessitating intervention by water service personnel for unclogging. Touho has six catchments, each serving between one to five tribes. Consistent with the approach adopted in Thio, three out of four water service personnel in Touho are indigenous Kanak residents of the commune. While acknowledging the necessity of charging residents for water consumption to finance pipeline maintenance and engage private service providers for technical tasks like chlorination or the installation of water reservoirs, they collectively assert the impracticality of such a system in Touho. This assertion stems from the potential obstruction of access to catchments by land-owning residents, who question the rationale for paying for water originating from their own land "After all, the water flows on our land, so why should we pay for water that comes from our own land?" This entrenched perception of water as inherently linked to customary land, fiercely claimed and fought by clans, persists strongly in Touho. Despite the predominant Kanak representation within municipal services, these services are perceived as operating outside the realm of customary logic. Consequently, although residents utilize the water supply service, the grappling with the idea of paying for its infrastructure remains a challenging prospect.

These tensions demonstrate that water payment, even when intended for the common good or communal collective purposes, is not accepted by the population of Touho; rather, it is perceived as a form of dispossession. These tensions extend beyond the issue of water, as they not only assert the population's resistance to what they perceive as an external intrusion but also allow them to reaffirm their ownership over land and associated resources, drawing a parallel with mining. There were initial mining explorations in the commune of Touho. However, local folklore recounts that the elders resisted any mining activity by magically concealing the mineral-rich soils of the commune making it impossible to find access to this site.

13.5.3 Poum: Containing, Separating and Accessing Water

Contrary to Touho, Poum located in the far-north of New Caledonia and including four inhabited and several uninhabited islands and islets, bears the hallmark of a long and sometimes (notably today[13]) chaotic history of nickel extraction (Budna, 2021). Actually, the (negative) interplay of water access, land degradation and nickel extraction started early and triggered conflict between local populations and companies. For instance, in 1962, a conflict broke out between SMSP (a mining company owned by a Caledonian political and economic entrepreneur, Jacques Lafleur), and the Titch tribe over mining waste dumped on their land. Lafleur had to pay cash compensation to the Titch chief, build a canal and deliver ten truckloads of sand.[14] Later, in 1967, the words of an elected council member sound like those of a "whistle-blower":

> M. Avallard—I would like to draw the Council's attention to the dangers that could arise from prospecting for nickel ore, which is currently taking place fairly close to the dam of the POUM water pipe (note that the first dam has already been completely damaged by mining discharges).
>
> It was decided to write to the prospecting company, warning them against these dangers and to remind them that there are limits to be respected on the perimeters of the water intakes.[15]

In 1978, the subject came again to the forefront, as illustrated by this extract from a letter from the Poum municipal council to the head of the subdivision:

> In times of high water, the streams flowing down from the mining plateau and into the sea carry various types of detritus, including mining waste. This is a major nuisance for most of Poum's population, and in particular for the "Brown, Ha Ho" families and the tribe, whose land and plantations are damaged by these muddy waters. The municipal council and I would like to see this problem tackled so that hygiene can reign in our village.[16]

Similar stories are reported in Thio in the same historical phase characterised by an enormous increase of extractive activities. These cases highlight the power of water in dis-enclaving mining and connecting it to surrounding spaces through

[13] Articles in les Nouvelles Caledoniennes newspaper.

[14] Carton 37 W69 Poum Lafleur, Fonds SLN, New Caledonia's Territorial Archives.

[15] Municipal archives of Koumac, Town council minutes. Ordinary meeting of first September 1967.

[16] Municipal archives of Koumac, Town council minutes. Ordinary meeting of sixth January 1978.

uncontrolled overflows—the negative or unmanaged side of a commons. Beside these forms of undesired connection, local people have to cope with water irregular supply, especially on the islets. Interviews show that they are ready to mix "traditional" subterranean local water and "modern" water coming from the main island, while keeping these two instances of water conceptually separated. Expressions such as "they don't have water but they have their water holes," the latter being sacred and surrounded by specific practices (only men can bathe there for instance), are no paradox but a sign of both the separation and dominance of modern water systems over traditional ones, even though both categories remain relevant. The connection of Yenghébane islet to the main island drinking water supply network (AEP) around 1995 led to the abandonment of the old collection basins. Thus, the ruined basin embodies the replacement of local water by modern water, one associated with the idea of development.

The association of modern water with development takes a specific shape in the context of a mining locality. Water access is a longstanding public problem in Poum as illustrated by its regular occurrence in *Les Nouvelles Caledoniennes* (the main daily newspaper in New Caledonia) and in the political debates for municipal polls. It became associated with mining development when SLN announced in 2019 the reopening of its extractive site in Thio and the handover of boreholes to the municipality. Water became entangled as a sort of bargaining chip in a mining development package based on a double promise of modernisation and emancipation, represented by the partnership between SLN, Poum municipality (ruled by pro-independence parties) and a local subcontracting company, SONAREP, owned by Kanak landowning clans (42%), the North Province investment agency SOFINOR (34%) and individual shareholders (24%).

13.6 Conclusion

Exploring NC's new shared water policy and the local-level politics of water in Thio and other settings in New Caledonia allows us to move away from over-simplistic dichotomies such as neo-liberal imaginary of governmentality and commodification vs the commons and discuss conflicting interpretations of water payment or what it means to "share" water, all the more with powerful neighbours and big water consumers (and polluters) such as mining companies. It is a way to "pose questions about the complicated relations between grand political economic considerations and the seemingly banal, taken-for-granted act of turning on a tap" (Ekers & Loftus, 2008: 713). There is of course a discrepancy between both levels of analysis as the new PEP is awaiting its implementation on the ground while local water policy has already—and for a long time—unfolded as a complex arena of interpretive contestations and political negotiations.

However lessons can be learnt from the local politics of water policy as it has been evolving in Thio, Touho or Poum. Lessons do not equate with formulas that could be applied mechanically. Actually, the main lesson we can draw from the case studies we carried out in different localities regards the entanglements of water policy in localised political and social issues. They are also rooted in the long history of settler colonialism and the often brutal ruptures it has introduced. The tendency toward segmentation in local politics and the multi-residential patterns reflecting the interplay of different historical layers of land dispossession, ethnic segregation and mining development play a key role in this respect.

The second lesson to be drawn—and once again, no dreaming of automatic formulas!—relates to individual, collective and distributed agency (Bennett, 2010; Wagner, 2013). The entanglements of governmentality, sovereignty, commodification and the commons, as regimes of discourse and practices, are shaped by these various forms of agency. Water assemblages result from the enactment, interpretation and transformation of these historical layering and sedimentation processes through different forms and levels of agency. These transformations reshape the composition of water-centered assemblages and especially how they include or exclude humans, land and material and immaterial nonhumans and thus generate differentiated ontologies.

The abstraction process resulting in the reduction of water as H2O is never achieved (Swyngedouw, 2005; Linton, 2010). It is both an ideological construct and a political horizon, with countervailing forces constantly at play, determining various forms of "ontological cleaving"—the "simultaneous action of drawing together and holding apart ontological multiplicities" (Lavau, 2013: 418) as shown in Poum. Such an analytical thread offers useful perspectives on emerging (and basically unavoidable) tensions in water policy and provides an understanding of the difficulties faced by managers, difficulties in changing practices or in creating institutional innovation, while providing avenues for productive negotiations involving all stakeholders, emphasising that this means also inviting nonhuman entities to the table.

References

Agrawal, A. (2005). *Environmentality: Technologies of government and the making of subjects*. Duke University Press.

Amin, A., & Howell, P. (Eds.). (2016). *Releasing the commons: Rethinking the futures of the commons*. Routledge.

Banks, G., & Le Meur, P.-Y. (2024). Closure and connection: A Southwest Pacific reappraisal of the mining enclave. In T. Rodon, S. Thériault, A. Keeling, S. Bouard, and A. Taylor (Eds.), *Mining and indigenous livelihoods: Rights, revenues, and resistance* (pp. 42–70). Routledge.

Bencivengo, Y. (2010). *La Société Le Nickel. Une entreprise au coeur de la naissancede l'industrie du nickel (1880–1914)*. PhD Dissertation, Université Paris 1 Panthéon-Sorbonne (3 volumes).

Bennett, J. (2010). *Vibrant matter. A political ecology of things*. Duke University Press.

Bouard, S. (2013). *La ruralité kanak à la recherche d'un modèle décolonisé*. Éditions Au vent des îles.

Bouard, S., Lejars, C., Coulange, D., & Sabinot, C. (2021). Gouvernance de l'eau sur terres coutumières en Nouvelle Calédonie: De la connaissance des savoirs locaux à la production d'outils de gestion adaptés et pertinents pour les acteurs locaux. In *Final research report, Goutte Project, CRESICA: Nouméa*.

Budna, M.-E. (2021). *L'eau plurielle, une résistante à l'enclavement dans une commune minière (Poum, Nouvelle-Calédonie)*. Master thesis in anthropology, EHESS.
Coulange, D., Bouard, S., Lejars, C., & Godin, P. (2020). L'eau douce en Nouvelle-Calédonie, continuités et ruptures d'une gouvernance hybride. In *Une étude de cas des vallées de la Pouanlotch et du haut de la Hienghène, Unpublished research report, Goutte Project*. IAC.
Dardot, P., & Laval, C. (2014). *Commun. Essai sur la Révolution au XXIe siècle [Commons: Essay on revolution in the twentieth century]*. La Découverte.
Dean, M. (2010). *Governmentality: Power and rule in modern society* (2nd ed.). Sage Publications.
Demmer, C. (2009). Secrets et organisation politique kanake. *Pour Sortir des Catégories Privé/ Public, L'Homme, 190*(2), 79–104.
Ekers, M., & Loftus, A. (2008). The power of water: Developing dialogues between Foucault and Gramsci. *Environment and Planning D: Society and Space, 26*(4), 698–718.
Ey, M., & Sherval, M. (2016). Exploring the minescape: Engaging with the complexity of the extractive sector. *Area, 48*(2), 176–182. https://doi.org/10.1111/area.12245
Fache, E., Le Meur, P.-Y., & Rodary, E. (2021). The new scramble for the Pacific: A frontier approach. *Pacific Affairs, 94*(1), 57–76.
Fernandez, S. (2014). Much ado about minimum flows… Unpacking indicators to reveal water politics. *Geoforum, 57*, 258–271.
Foucault, M. (1982). Afterword: The subject and power. In H. Dreyfus & P. Rabinow (Eds.), *Michel Foucault: Beyond structuralism and hermeneutics* (pp. 208–226). University of Chicago Press.
Foucault, M. (2002). Power. In J. D. Faubion (Ed.), (R. Hurley et al., Trans.) *Essential works of Foucault 1954–1984*. Penguin Books.
Foucault, M. (2004). *Sécurité, territoire, population, cours au collège de France (1977–1978)*. Gallimard-Seuil.
Gosset, L. (2020). Les lieux d'eau remarquables de la tribu de Saint-Paul, Thio. In *Research Report, Noumea, IRD, Goutte Project*.
Hardt, M., & Negri, T. (2009). *Commonwealth*. Harvard University Press.
Hartog, F. (2003). Régimes d'historicité. In *Présentisme et expériences du temps*. Seuil.
Herrenschmidt, J.-B., & Le Meur, P.-Y. (2016). *Politique foncière et dynamiques coutumières en Nouvelle-Calédonie et dans le Pacifique*. Province Nord-IRD.
Lavau, S. (2013). Going with the flow: Sustainable water management as ontological cleaving. *Environment and Planning D: Society and Space, 31*, 416–433. https://doi.org/10.1068/d25411
Le Meur, P.-Y. (2013). Locality, mobility and governmentality in colonial/postcolonial New Caledonia. The case of the Kouare tribe (xûâ Xârâgwii), Thio (Cöö). *Oceania, 83*(2), 130–146.
Le Meur, P.-Y. (2017). Le destin commun en Nouvelle-Calédonie: Entre projet national, patrimoine minier et désarticulations historiques. *Mouvements, 91*(3), 35–45.
Le Meur, P.-Y. (2024). Mining fronts, labor mobilities, and the construction of locality in Thio, New Caledonia. In G. Castillo, D. Brereton, & M. Himley (Eds.), *Mining, mobility, and social change in the global south. Regional perspectives* (pp. 165–185). Routledge.
Le Meur, P.-Y., Levacher, C., Bouard, S., Herrenschmidt, J.-B., & Sabinot, C. (2021). Mining and the value of place in New Caledonia: Negotiation, evaluation, recognition. *The Extractive Industries and Society, 8*, 44–54.
Lejars, C., Bouard, S., & Ferrand, N. (2021). La politique de l'eau partagée en Nouvelle-Calédonie: Retour d'expériences sur un dispositif de co-construction et de co-planification. *Sciences, Eaux & Territoires, 35*, 68–75.
Levacher, C., & Le Meur, P.-Y. (2022). The compensation arenas in South New Caledonia. Minescape management, governmentality and politics. *The Extractive Industries and Society, 11*, 100999. https://doi.org/10.1016/j.exis.2021.100999
Li, T. M. (2014). What is land? Assembling a resource for global investment. *Transactions of the Institute of British Geographers, 39*, 589–602.
Linton, J. (2010). *What is water? The history of a modern abstraction*. University of British Columbia Press.

Linton, J., & Budds, J. (2014). The hydro-social cycle: Defining and mobilizing a relational-dialectical approach to water. *Geoforum, 57*, 170–180.
Moore, S. F. (1987). Explaining the present: Theoretical dilemmas in processual ethnography. *American Ethnologist, 14*(4), 727–736.
Orlove, B., & Caton, S. C. (2010). Water sustainability: Anthropological approaches and prospects. *Annual Review of Anthropology, 39*, 401–415.
Ostrom, E. (1990). *Governing the commons. The evolutions of institutions for collective action*. Cambridge University Press.
Peytavi, O. (2024). Quand la mine coupe l'eau: Mémoires d'un flux brisé. L'Affaire N'Goye à Petit Borendy, Thio (Nouvelle-Calédonie). *Journal de la Société des Océanistes, 35*, 158–159.
Peytavi, O., Bouard, S., Le Meur, P.-Y., & Lejars, C. (2023). Freshwater supply as sociotechnical tinkering: The co-creation of water knowledge and assemblages in New Caledonia. *Journal of Political Ecology, 30*(1), 5289. https://doi.org/10.2458/jpe.5289
Richard, D., Sabinot, C., Gosset, L., Worliczek, E., Pidjo, A., Bertrand, M., & Garcin, M. (2021). Towards a better integration of local populations in the design of projects to manage the massive aggradation of rivers stemming from mining activity in Thio, New Caledonia. *The Extractive Industries and Society, 8*, 64–73.
Rose, N. (1999). *Powers of freedom: Reframing political thought*. Cambridge University Press.
Strang, V. (2005). Commons senses. Water, sensory experience and the generation of meaning. *Journal of Material Culture, 10*(1), 92–120. https://doi.org/10.1177/1359183505050096
Strang, V. (2006). Introduction. Fluidscapes: Water, identity and the senses. *Worldviews, 10*(2), 147–154.
Strang, V. (2009). *Gardening the world: Agency, identity and the ownership of water*. Berghahn.
Swyngedouw, E. (2005). Dispossessing H2O: The contested terrain of water privatization. *Capitalism Nature Socialism, 16*(1), 81–98. https://doi.org/10.1080/1045575052000335384
Swyngedouw, E. (2015). *Liquid power. Contested hydro-modernities in twentieth-century Spain*. MIT Press.
Trépied, B. (2011). Des conduites d'eau pour les tribus. Action municipale, colonisation et citoyenneté en Nouvelle-Calédonie. *Revue d'Histoire Moderne et Contemporaine, 58*(4), 93–120.
Wagner, J. R. (2012). Water and the commons imaginary. *Current Anthropology, 53*(5), 617–641. https://doi.org/10.1086/667622
Wagner, J. R. (Ed.). (2013). *The social life of water*. Berghahn Books.

Policy Documents

Forum H2O. (2018). L'eau, un enjeu stratégique essentiel. Press Release, Nouméa, 18–20 April 2018.
Government of New Caledonia. (2022a). *Loi du pays relative au domaine public de l'eau de la Nouvelle-Calédonie et à la protection de la ressource en eau*. Government of New Caledonia.
Government of New Caledonia. (2022b). *Rapport de présentation de l'avant-Projet de la loi du pays relative au domaine public de l'eau de la Nouvelle-Calédonie et à la protection de la ressource en eau*. Government of New Caledonia.
ISEE. (2019). *Municipal statistics*. ISEE.
PEP. (2019). *Schéma d'orientation pour une politique de l'eau partagée*. Government of New Caledonia.

Chapter 14
Towards New Forms of Water Governance to Improve Indigenous Inclusion

Séverine Bouard, Caroline Lejars, Pierre-Louis Mayaux, Sylvain Barone, and Adèle Bommier

Abstract One of the first objectives of the new Caledonian water policy is the implementation of a new governance system. This chapter explores possible new governance rules for New Caledonia. It presents different scenarios ranging from the status quo (conservation of the current system, with an inter-service management that coordinates provinces and municipalities, but led by the government), to the establishment of a single water office (to pool resources and skills), a water agency (based on the French model) or new forms of governance including public representatives. It focuses on the issues raised by the inclusion of indigenousness in water governance arrangements in order to provide perspectives for New Caledonian practitioners in the formalisation of their own governance model. This topic aligns with the broader theme of the incorporating indigenous claims into public social,

S. Bouard (✉)
IAC (New Caledonian Institute of Agronomy), Pouembout, New Caledonia

LU, Lincoln University, Lincoln, New Zealand
e-mail: severine.bouard@iac.nc

C. Lejars
CIRAD (Agricultural Research Centre for International Development), UMR GEAU, Montpellier, France

CIRAD (French Agricultural Research Centre for International Development), UMR GEAU (Univ Montpellier, CIRAD, INRAE, IRD, BRGM, AgroParisTech, Institut Agro), Montpellier, France

P.-L. Mayaux
CIRAD (Agricultural Research Centre for International Development), UMR GEAU, Montpellier, France

S. Barone
INRAE (Research Institute for Agriculture, Food and Environment), UMR G-EAU, Univ Montpellier, AgroParisTech, CIRAD, Institut Agro, IRD, Montpellier, France

A. Bommier
CIRAD (Agricultural Research Center For International Development), UMR GEAU, Univ Montpellier, IRD, CIRAD, INRAE, BRGM, Institut Agro, AgroParisTech, Montpellier, France

DGTM, General Directorate of Territories and the Sea of Guyane, Cayenne, Guyane

C. Lejars, S. Bouard (eds.), *Water Policy in New Caledonia*, Global Issues in Water Policy 32, https://doi.org/10.1007/978-3-031-88196-1_14

economic and environmental policies affecting indigenous territories. The comparative study of the Australian, Canadian and Bolivian cases makes it possible to identify issues regarding the inclusion of indigenous peoples as an entity in its own right, and the integration of the cultural norms, values and practices of these peoples into the mechanisms and public policies as a result. This comparison therefore provides an opportunity to suggest key considerations regarding international concepts and water governance models, as well highlighting New Caledonia's encouraging progress in integrating indigenousness within the water governance system.

Keywords Water governance · Indigenousness · Comparison

14.1 Introduction

In the international context of promoting indigenous identities and seeking methods for taking into account indigenous knowledge on sustainable ecosystem management, this chapter focuses on the integration of indigenousness into water governance arrangements. Water-related issues are among the most significant challenges faced by indigenous peoples. They affect the notions of ownership of lands and resources, access to economic development, protection of their vital ecosystem and, above all, their recognition as a people and the recognition of their water uses and customs by the various state authorities. Through a comparative study of selected cases, this chapter examines the issues raised by the inclusion of indigenousness in governance mechanisms and the development of approaches for the practical integration of these issues. The objectives of this chapter are therefore twofold: to identify specific issues raised by the consideration of indigenousness in governance mechanisms through three case studies (Australia, Canada and Bolivia) and to propose perspectives for both researchers and New Caledonian water practitioners.

14.2 Consideration of Indigenous Peoples in Water Governance Principles

14.2.1 Principles of Water Governance: The OECD Framework

For the Organisation for Economic Co-operation and Development (OECD), water governance is the set of "political, institutional and administrative rules, practices and processes (formal and informal), customs and beliefs, through which decisions are made and implemented, stakeholders express their interests and these are taken into account, and decision-makers are held accountable for the management of water resources and water utilities" (OECD, 2015a, b).

It is a dynamic process that incorporates the interplay of power and influence between stakeholders and refers to decision-making on water in relation to sectors of activity, from local to international. This procedural and iterative dimension means that decisions may be regularly adapted and adjusted according to the effects produced by the system in place.

The governance system refers to different organisations with a high degree of interdependence in managing water and forms the institutional framework established by the governing authority. These are the formal rules that may be laid down in a law, for example. The governance system also includes all the informal and practical rules that can be shared within a working group to produce public policies. This governance system, made up of rules, mechanisms and instruments, enables the design of public policies that are effective, efficient, trustworthy, and backed by strong engagement. These three effects are broken down in the OECD's 12 principles of water governance (Fig. 14.1), which are designed to ensure "good" governance.

For Jiménez et al. (2020), the governance system brings together a set of functions (such as planning, financing, monitoring, standards and rules, and capacity development) and qualities such as being multi-level, inclusive and adaptive. It is the combination of these two aspects, underpinned by the values and aspirations of the actors involved, that makes it possible to observe results such as changes in the behaviour of stakeholders and citizens, and the sustainability and resilience of the system. The governance system is not just a set of rules, but a process combining functions and qualities that produce results.

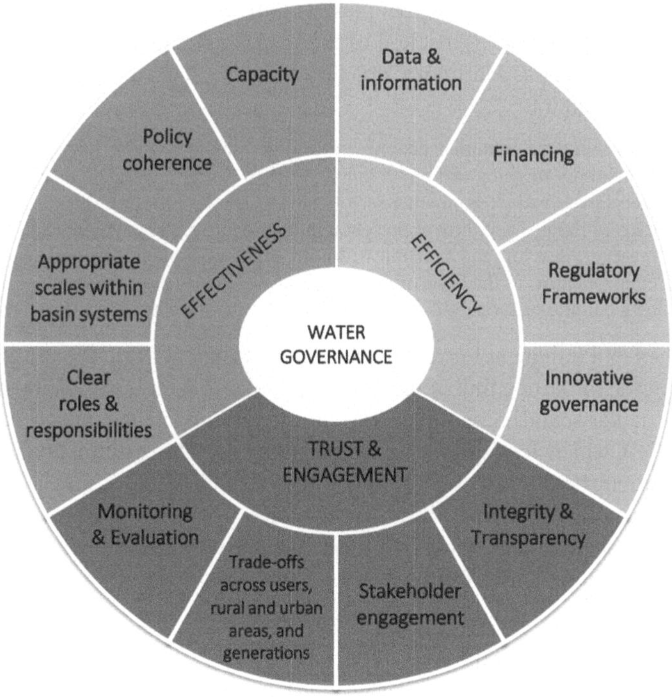

OECD principles on water governance (2015)

Fig. 14.1 OECD principles on water governance

In a context of legal and cultural pluralism,[1] as in New Caledonia, the challenge is to build a system of governance that takes various issues into account in policy decisions—including policy coherence, commitment and fair representation of both indigenous and non-indigenous stakeholders, as well as the inclusion of heterogenous forms of knowledge and perspectives.

Nonetheless, the cyclical approach to water governance also promoted by OECD—Policy design and strategy, implementation, monitoring and evaluation in a loop—rarely stands up to sociological analysis of the process being promoted (Lascoumes & Le Galès, 2012; Hassenteufel, 2011). These sociological and anthropological approaches are essential to integrating indigenous values or representations of water into the governance system.

14.2.2 Concepts to Be Integrated into Water Governance: Indigenousness, Minorities

14.2.2.1 Indigenousness vs. Minority

Indigenousness is often associated with notions of minority and self-determination or sovereignty over natural resources.

Although there is no single authoritative definition, there have been various criteria to define indigenous peoples. The main one is the "sense of belonging to an indigenous people" (Article 2 of the ILO Convention). Beyond that, the definition proposed by José Martinez Cobo (Taylor et al., 2019) is one of the most widely used in the literature:

> Indigenous communities, peoples and nations are those which, having a historical continuity with pre-invasion and pre-colonial societies that developed on their territories, consider themselves distinct from other sectors of the societies now prevailing in those territories, or parts of them. They form at present non-dominant sectors of society and are determined to preserve, develop and transmit to future generations their ancestral territories, and their ethnic identity, as the basis of their continued existence as peoples, in accordance with their own cultural patterns, social institutions and legal systems. (José Martinez Cobo[2])

The UN Permanent Forum on Indigenous Issues (UNPFII) has added the following criteria: the strong link of indigenous peoples to their territory and natural resources; particular social, economic and political systems: a language, a culture and a set of beliefs which form the basis of their identity. The specific connection to the land distinguishes them from minorities, even if some of their rights are similar (European Convention on Human Rights on the Rights of Minorities, 1993). This is the main difference that has a major impact on the way these populations are taken into account in systems of governance.

---

[1] See Chap. 5 and Part II.

[2] Martinez Cobo, J. 1986/7. Study of the Problems of Discrimination against Indigenous Populations. UN Doc. E/CN.4/Sub.2/1986/7 and Add. 1–4. paras 379–383.

The 2007 United Nations Declaration on the Rights of Indigenous Peoples (UNDRIP) is not an internationally binding instrument, but it does define rights and freedoms, primarily the right to self-determination (Taylor et al., 2019). Its article 3 states that indigenous peoples "[...] freely determine their political status and freely pursue their economic, social and cultural development" (Taylor et al., 2019).

Self-determination is one of the key demands of indigenous peoples concerning their ability to freely determine their future, and in particular their participation in governance arrangements proposed or imposed by states. Article 5 emphasises the need to recognise customary systems of organisation. Article 25 goes further, recognising the right of indigenous peoples to maintain and strengthen their spiritual and special ties with the territories and resources they occupy, live in and use.

This right is of particular interest when considering the diverse ways water resources are used and the protection of these customary practices. Article 32 also gives indigenous peoples the right to establish priorities and strategies for the development and use of their lands and other resources. As any project has an impact on indigenous peoples' resources, they must at least be consulted and give their free, prior and informed consent to the project.

Taylor et al. (2019) conclude that the establishment of institutions or processes that enable this directive to be put into practice is essential to the construction of water resource governance arrangements.

On the basis of these observations and definitions, how can we link water governance and indigenousness, and how can we take into account, from a procedural point of view, the norms, values and representations of the Kanak people of New Caledonia? What can we learn from experiences in Canada, New Zealand and Australia?

There are different ways of taking indigenousness into account. In this context, taking into account means including indigenous peoples as a group of actors in governance mechanisms, and by indigenousness we mean the inclusion of indigenous peoples as an entity in their own right and as bearers of their own values, norms, representations and practices that constitute their identity.

14.2.2.2 Taking into Account the Specificities of Indigenous Actors in Relation to the Majority System

The main issue is whether indigenous peoples are considered as one of the stakeholders on an equal footing with others, or whether they are considered as an entity with their own rights and a privileged place, or at least different from the other interest groups in the commission (von der Porten & de Loë, 2014a, b). Promoting governance arrangements that allow all types of stakeholders to interact to find solutions is a widespread idea in water policy, especially within the paradigm of integrated water resources management. However, when indigenous peoples are treated in the same way as a representative of an environmental protection association on a basin governance committee, they lose some or all of their ability to assert their rights, particularly the right to self-determination. Indigenous peoples then run the risk of facing the same difficulties as minority actors whose interests are difficult to reconcile with the majority on the committee, and the decisions taken by the committee may not include their claims.

14.2.2.3 Respecting Indigenous Identities: Values and Norms

How can governance processes allow for the indigenous values associated with a river to be taken into account? How can the spiritual nature of water, the symbolism of the cycle of life associated with its flow from source to sea, the link between people and their ancestors, and its purity be taken into account in the actions taken to protect the river from pollution and preserve its ecological continuity?

This type of questioning has been particularly well expressed in the case of the Whanganui River in New Zealand (Hsiao, 2012). As the statement "I am the river and the river is me"[3] attests. In this context, taking into account the values Maori people associate with the river is taking into account the people themselves.

The idea of taking indigenous values into account is never far from the idea of indigenous people as social actors in governance. When both considerations are advanced together, indigenous people participate in the decision-making process and indigenous knowledge, social representations and values are incorporated into public policy.

14.2.3 Taking Account of Indigenous Identities: Integration, Inclusion or Incorporation of Knowledge and Practices?

Indigenous knowledge and practices can be integrated in different ways. The term 'inclusion' implies the existence of cross-fertilisation or hybridisation mechanisms (David & Lecercle, 2020) between the knowledge, worldviews and methods of the customary world and those of non-customary people. Hybridisation can take place in a legal, legislative and normative system, in an action program or a public policy. The notion of inclusion is not exclusive of notions of legal pluralism. In most cases, legal pluralism is the result of cultural pluralism, in which legal rules from different legal systems coexist, sometimes overlap and are applied differently depending on the civil status of the person within the same state and for identical legal situations (David & Lecercle, 2020).

Where there is recognition of a right in addition to ordinary law, customary law is then incorporated into the state's more global system of norms, with entanglements of rights at different levels. Recognised customary law may, for example, only deal with matters of civil rights (Cornut & Deumier, 2016), with ordinary law

[3] Whanganui Iwi and The Crown, Tūtohu Whakatupua, 30 August 2012, at para. 1.1.

applying to other areas of law. The notion of inclusion recognises the reciprocal influences that one value system can have on another. The incorporation of customary norms into so-called common law norms can have effects on these customary norms (weakening or strengthening) as well as on common law norms (confirmation, annulment). Conversely, the term 'incorporation' implies a process of coercion or an *a posteriori* integration of one system into another in a unidirectional manner, without mutual adaptation. If indigenousness is taken into account only through the participation and consultation of the indigenous peoples concerned, without co-learning and co-building processes, we are likely to see the incorporation of indigenous norms, values and representations into the pre-established state system. There is no adaptation to the methods or vision of the mainstream world, which itself must be integrated a posteriori.

Each people group has its own concept of water, with particular values based on associated rituals or traditions. But the majority of indigenous peoples have a special attachment to this resource because it is linked to their identity. It symbolises their connection to place and land, supports their customs and beliefs, and is an integral part of their traditional knowledge (Jackson & Altman, 2009).

A key focus is the impact on ecosystem functioning, where relationships between non-human actors and humans are crucial. Indigenous peoples have a holistic view of the environment (Durette, 2010a). For example, the distinction made in French law between water as the 'common heritage of the nation', and riverbeds which are the property of the state, local authorities or private individuals up to half the length of the bed, has no equivalent in the New Caledonian system of norms (Drobenko, 2013). Article 44 of the Organic Law of New Caledonia refers to "watercourses and lakes", meaning both the water and the bed that carries it (Massenavette, 2011). In other words, in the Organic Law of New Caledonia, land and water are considered in the same way, in accordance with customary representations. This is one of the first steps towards incorporation.

Nevertheless, the status of water is still an issue. This holistic view of the environment and connection to the land is one of the main characteristics of indigenous peoples, with reference to Article 25 of the UNDRIP, with its 'special spiritual ties to coastal lands, territories and marine areas'. One of the notable differences between indigenous and non-indigenous peoples is that for indigenous peoples, 'water' is not an inert element or a 'commodity', but has a relationship with people (Taylor et al., 2019). The term 'water-related uses' should therefore not be limited to domestic, economic and environmental uses, but should also include uses for traditional activities, culture spiritual life and recognition of the connection to a resource that has immeasurable value and no utilitarian purpose (Finn & Jackson, 2011).

Indigenous peoples have very detailed and comprehensive knowledge of the ecosystem around them, passed down from generation to generation. Arsenault defines the term 'traditional knowledge' as the set of practices and beliefs adapted to the local culture and environment, and the relationships between living beings and with their environment, that have lasted for centuries (Arsenault et al., 2018). Indigenous knowledge includes both scientific and spiritual elements.

14.3 Inclusive Water Governance: Lessons from Australia, Canada and Bolivia

Since 2007 and the adoption of the UNDRIP, research on the inclusion of indigenous peoples and norms in water governance models has increased. Research in Australia, Canada, Bolivia and New Zealand is particularly instructive for New Caledonia, as similar issues and comparable contexts exist for discussing the possibilities of implementing more inclusive water governance. In addition, these countries offer substantial literature addressing issues of water governance, natural resource management or protected area management in relation to the presence of indigenous peoples. These research articles testify to the interest of the actions undertaken in these countries, particularly with regard to innovations in governance arrangements or legal systems, and have enabled us to document the comparison in detail.

14.3.1 Introduction to the Specific Features of Each Case Study

14.3.1.1 Australia and the Murray-Darling Basin

In addition to the diversity of Indigenous communities in Australia, there is also a diversity of laws in each territory, with regulations that are more or less supportive of Indigenous recognition (David & Lecercle, 2020). In 2016, Indigenous people made up 3.3% of Australia's population, with the exception of the Northern Territory, where Indigenous people make up almost 30% of the population. At the federal level, the 1993 Native Title Act allows Indigenous people to apply for recognition of their rights to land and water resources. This title allows them to engage in subsistence fishing without the need for special permits or other restrictions imposed elsewhere (Jackson et al., 2012). However, people's ability to have their native title recognised is relative: of the 1890 applications, 474 were considered valid and only 394 were eventually recognised. This low level of title recognition reflects the difficulty governments have in recognising these communities (David & Lecercle, 2020). In 2004, an agreement was signed between the Australian government, states and territories: the National Water Initiative. In theory, this agreement requires local authorities to consider the water needs of communities in their governance arrangements, but it does not set out methods to facilitate this consideration in management plans. In practice, initiatives to consider Indigenous claims vary widely between states and local governments. The Northern Territory and Queensland provide fertile ground for comparative studies of collaborative water management practices that can shed light on barriers or factors that facilitate Aboriginal inclusion in plans

(Jackson & Altman, 2009; Holley & Sinclair, 2014; Hughey et al., 2017; Jackson et al., 2019; Poelina et al., 2019).

This work has also focused on research initiatives based on Indigenous methods and their implications for the (ac)knowledge of Indigenous values in governance arrangements (Hill et al., 2012; Woodward et al., 2012). The Murray-Darling Basin has been studied extensively in relation to the 2007 Water Act. In this example, 'cultural flows', a legal and legislative tool designed in this basin to recognise Indigenous non-commercial and traditional uses in a way that complements environmental flows, is particularly inspiring (Cranney & Tan, 2011; Finn & Jackson, 2011; Durette, 2010b).

14.3.1.2 Canada and the Province of British Columbia

Canada offers stimulating comparisons with New Caledonia. The issues surrounding water and indigenous peoples are similar, particularly in relation to mining, the pollution it causes, the degradation of their vital environment and the difficulty of accessing good quality drinking water. In addition, power is fragmented across federal, provincial and municipal levels, which leaves a wide scope for the application of laws between the different provinces, especially on the issue of indigenous peoples. Land, subsoil and natural resources are part of the public domain of the provinces, which has a major impact on indigenous peoples' access to these resources in their territories (Cardenas, 2007). In 2016, the proportion of Indigenous peoples was estimated at 5% of the total population (of which First Nations 65%, Métis 30% and Inuit 5%). Canada's indigenous peoples are recognised in the 1982 Canadian Constitution as having special rights by virtue of their aboriginal status. In the province of British Columbia, work on collaborative governance in an Indigenous context has been undertaken in line with the province's strategy to include this type of mechanism in its reform of water legislation (Bradford et al., 2016; von der Porten & de Loë, 2013b). This work has made it possible to identify a number of favourable and unfavourable factors for the practical inclusion of Indigenous peoples as stakeholders in the systems, while preserving their capacity for self-determination (Simms et al., 2016; von der Porten & de Loë, 2013a, 2014a, b). Furthermore, for the people of this province, water is an essential element of their spirituality and identity, as are the aquatic environment and the fish products that form the core of their diet (von der Porten & de Loë, 2013b).

14.3.1.3 The Bolivian Plurinational State and Its Constitution

A similar country, but less well known to New Caledonians, Bolivia has an innovative constitution that places indigenous peoples at the centre of its framework (Canessa, 2014). Demographically, the indigenous population represented almost

50% of the population in 2017 (David & Lecercle, 2020). The various indigenous communities are highly structured and have been strongly represented in the political arena since the great indigenous mobilisation and struggle of the 1970s until the adoption of a new constitution under Evo Morales in 2009. There is strong pressure on indigenous territories, particularly from deforestation and the exploitation of mining resources (Boelens et al., 2007). As described by Lacroix, the Constitution of the Plurinational State of Bolivia recognises the fundamental rights of indigenous peoples in line with the United Nations Declaration on the Rights of Indigenous Peoples (UNDRIP) (Lacroix, 2013) and, for example, recognises 37 official languages. In addition, a specific law on the protection of the rights of nature was adopted in 2010. These legal innovations tend to promote the inclusion of indigenous peoples and their values in governance mechanisms. However, the reality is more nuanced, spanning a spectrum from scenarios that support community self-determination to those where communities are dominated by municipalities (Tockman et al., 2015). What is interesting in some regions, however, is the indigenous organisation for the provision of community water services, self-organised as a committee, as in Cochabamba. This type of initiative raises the question of integrated management between these indigenous community bodies and local or state governments (David & Lecercle, 2020), but for our purposes it is above all Bolivia's constitutional innovation that provides food for thought on how to integrate indigenous peoples into governance arrangements.

14.3.2 The Challenges of Taking Indigenousness into Account in Water Governance

A detailed analysis of the literature on these different case studies has identified five types of issues that cannot be ignored when considering Indigenous peoples and/or their customs in Indigenous governance arrangements.

To present these issues, we use as a reference article the work of Jackson and his team in 2019 (Jackson et al., 2019) with Indigenous communities in Northern and Western Australia. Supplemented by the findings of the work of Jackson, 2018, Escott et al., 2015, Jackson et al., 2012, Maclean, 2015 and Simms et al., 2016, we can identify five types of issues related to the consideration of indigenousness in water governance: (1) the governance and its modalities, (2) cultural values and norms, (3) economic and financial aspects, (4) capacity and skills, (5) data and information. In fact, what comes under the heading of governance methods and indigenous cultural norms and values are the main factors of variation in the way indigenousness is taken into account in a governance system. The other issues discussed appear consistently in the literature, where authors propose explanations for the varying levels of inclusion of indigenous peoples in a given territory. However, they appear as indirect factors because they result from the integration of the previous issues.

14.3.2.1 Governance System and Procedures

One of the first difficulties in hybridising Western and indigenous governance is the compatibility of methods used with indigenous practices in the governance process. In Australia, Escott and Jackson observe a gap between the bureaucratic practices of the states and those of Indigenous peoples. These processes, which are heavily framed by administrative and technocratic procedures, are not adapted to indigenous practices. Protocols for invoking the spirits of the land, or the time needed to reach a decision in order to gain the consensus or consent of community members, are not always respected.

In addition, non-indigenous people may find it difficult to recognise customary organisations and their methods of appointing representatives. These methods are sometimes at odds with the values of democracy and gender and intergenerational equity. In such cases, non-indigenous people must either accept these methods of appointment outright or find a compromise to elect representatives based on democratic values. There is a risk of creating conflict within communities over the legitimacy of the representatives chosen, or, on the contrary, of benefiting by thwarting mechanisms of corruption or the marginalisation of more vulnerable groups within indigenous peoples.

The question of legitimacy can also be challenged by the scale or the size of the governance territories finally chosen, is it appropriate and does it match with the indigenous territories? Simms et al. (2016) highlight the difficulties posed by the scales considered for water governance in Australia. The administrative (states, provinces and municipalities) or hydrographic (catchment) scales, for which institutional actors are well identified, do not necessarily coincide with Indigenous territories. This raises issue of the legitimacy of the people chosen to represent communities on these bodies.

Sometimes, communities refuse to sit on these bodies because they consider that these watershed organisations deny them their right to self-determination by forcing them to conform to the structure in place. In Canada, Tarte (2015) notes that indigenous representatives sit on the participation bodies of only four of the nine watershed organisations studied. For indigenous representatives, the existing system prevents them from exercising and asserting their ancestral rights over their land, which is an integral part of their identity.

14.3.2.2 Cultural Norms and Values

The system of governance promoted by Western institutions has difficulty in recognising indigenous systems of values, practices and organisations at the same level as Western representations. This lack of mutual recognition sometimes hinders the inclusion of Indigenous peoples. In Australia, Jackson uses the example of water "waste". The Western injunction to turn off the tap clashes with Indigenous beliefs that see water as a cyclical element, leading to misunderstandings about Indigenous people's desire to preserve water (see also Chap. 6). Western governance frameworks end up more or less implicitly excluding indigenous systems of norms and knowledge (Simms et al., 2016).

Escott has shown that encouraging indigenous participation in both the decision-making process and the implementation of public policies depends on non-indigenous people recognising indigenous management systems and highlighting their practices (Escott et al., 2015). The difficulty is compounded by the need to adapt this official recognition to the diversity of communities. In a state that is home to several indigenous nations, each with their own systems, values and interests to defend (Jackson et al., 2012). This adaptation is difficult, as Jackson describes a tendency among practitioners to view indigenous peoples' participation solely in terms of their 'cultural values' or the protection of sacred sites, when this is not the only claim made by indigenous peoples.

14.3.2.3 Economic and Financial Issues

In Australia and Canada, Jackson describes the underfunding of maintenance and operation of water infrastructures for indigenous populations. Water networks are often outdated or insufficiently developed, making it challenging to supply drinking water to the most remote communities. A commitment to truly collaborative and adaptive governance would ensure funding for both infrastructure maintenance and leadership in multi-stakeholder forums. Also, the role of local facilitators is essential for the model's inclusion and adaptation, particularly in systems designed to decentralise decision-making and bring it as close as possible to the stakeholders in a given area. However, it is very costly.

One option would be to work with local authorities to explore financial mechanisms to support the system. For example, the concepts of royalties or taxes based on the service provided can be discussed in relation to the concepts of exchange or gift giving. Once financial resources have been consolidated, the question of autonomy in budget use by local actors remains and must be both respected and controlled in order to avoid misappropriation of funds.

14.3.2.4 Acknowledgment of the Diversity of Skills and Competencies

The skills specific to water management that are required to fully contribute to the governance process are a major obstacle to the inclusion of Indigenous peoples. The Canadian and Australian examples show that Indigenous people often have only a partial command of the technical vocabulary and know-how needed to assert themselves as leaders in project participation bodies.

Escott proposes developing training courses and creating employment opportunities in the water sector to familiarise indigenous people with the issues and methods used by non-indigenous actors. In this case, we are dealing with a non-reciprocal adaptation of indigenous methods and skills to those of non-indigenous people, with a significant asymmetry of information. Clearly, a middle ground must be found to foster a more reciprocal acquisition of knowledge and understanding between both parties.

14.3.2.5 Data and Information: Access to and Collection of Data on Indigenous Territories

In the vast majority of cases, there are gaps in information about communities' water resources and how they are managed. Without knowledge of their real needs or the exact cost of providing services, there is little visibility to adopt effective policies while respecting indigenous claims (Jackson et al., 2019).

Information asymmetries recur between indigenous peoples and water professionals, but also between forms of knowledge themselves, where a more or less implicit hierarchy is posited between scientific, empirical, cultural and traditional knowledge. The sacred or secret dimension of some indigenous knowledge poses an additional challenge when governance mechanisms aim to increase transparency, mutual cooperation and the sharing of strategic information.

This summary illustrates the difficulty of integrating indigenous issues into water governance, but Jackson goes even further. For him, the difficulty also relates to the risk of involving indigenous peoples in institutions where there is no real power-sharing in decision-making. In a system that is asymmetrical in terms of recognition of issues, knowledge and prerogatives, asymmetric participation can unintentionally revive strong political tensions associated with water governance. As a result, the governance system runs the risk of embedding a degree of mistrust between indigenous and non-indigenous actors (Jackson et al., 2019).

14.4 How Is Indigenousness Taken into Account in New Caledonia? Strengths and Challenges

How should the coexistence of different land tenure systems, norms, values and organisations in New Caledonia be approached? The question is how to take indigenousness into account in the water management system and, more generally, how to take cultural diversity into account in the governance of New Caledonia.

14.4.1 Assets of Governance Systems and Procedures

There are many water-related issues in New Caledonia, but the indigenous issue is not really among the ones identified by local actors as a barrier to the implementation of the Shared Water Policy. The issues that are perceived as barriers are more related to the distribution of powers, funding, human resources and lack of communication between different levels of government in carrying out their tasks. This issue of role distribution between the provinces and the government, which was partially resolved in 2020, is still the subject of debate due to the acceleration of budgetary cuts that New Caledonia has been experiencing since 2019.

The organic law is cited as an official reference for the distribution of roles, but in practice the institutional actors worked through informal arrangements for the implementation or non-implementation of actions (Pittet, 2019). Today, the delegation of responsibility for water management in New Caledonia has ceased. After many years of doing so, the provinces now only intervene in the water sector to authorise water abstraction in response to economic and demographic growth in their territory.

Having examined the issues of division of responsibilities between institutional levels, it is clear that the recognition and inclusion of values, practices, knowledge and legitimacies remains fundamental to the construction of a sustainable governance system. This is all the more true at the time of writing, in May 2024, when New Caledonia is facing its fourth 'Kanak revolt', with the indigenous people expressing their fear of becoming a minority in their own country and demanding full sovereignty. Whatever mechanism is put in place, the instrument used to implement the shared water policy must be designed to propose effective solutions that make it possible to achieve the policy's objectives and principles.

However, New Caledonia's particularity lies in its organisation, which is relatively favourable to the inclusion of indigenous peoples in general, and in the current context, there is an opportunity put this integration into practice through the creation of a water management system. The Kanak people and their identity are included and recognised as a people in the Organic Law; they are represented in New Caledonia's institutions through the Customary Senate; and they can use other channels to defend their interests in the design and implementation of policies, including partisan, political and bureaucratic channels. However, this institutional and official integration is carried out using management frameworks and methods that are not adapted to the unique organisational structure of the Kanak people. Despite this observation, the integration of the Kanak people into New Caledonia's governance and decision-making systems in general is relatively high, and New Caledonia has the potential to innovate in the practical implementation of a unique water governance system.

14.4.2 Integrating Knowledge of Indigenous Values and Social Organisation

Compared to the other case studies in Australia, Bolivia and Canada, New Caledonia is relatively advanced on the issue of taking indigenous peoples into account.

In order to take indigenousness into account, it is essential to understand and share the value and organisational systems of stakeholders, whether customary or not. This allows for clarity around claims and interests of stakeholders in relation to territories and to better understand the multi-level organisation of each party. In New Caledonia, the administrative organisation is complex and described in legal texts, while the Kanak organisation is sometimes difficult to identify (Dégremont & Sabinot, 2020). In the Loyalty Islands, the customary context was less disrupted by colonisation, which facilitated relations between chiefdoms, communes and the province. Institutions there follow the customary organisation, as in Bolivia, where a stable customary organisation facilitates the integration of indigenous people into politics. In New Caledonia, the integration of customary and modern systems is unique and still evolving. The island province is working to restore the ancestral system of Kanak chiefs to strengthen their legitimacy and representativeness. This process, which began in the 1990s, aims to recognise the chiefdoms as customary communities. However, this restructuring is more delicate on the mainland, where there is surface water to manage, but also where cultural and land pluralism is more pronounced, and where colonisation has profoundly altered the organisation of indigenous society, particularly as a result of population movements and land redistribution issues.

At the same time, work in the human and social sciences is being carried out to develop knowledge of changes in Kanak social organisation and their specific links to water resources (Bouard et al., 2020; Coulange et al., 2021; Chaps. 6–9). This new area of research helps understand and articulate the different value systems that need to be incorporated into management plans. However, as shown by Simms et al. (2016) in Australia, the administrative (states, provinces and municipalities) or hydrographic (catchment) scales, for which institutional actors are well identified, do not necessarily coincide with Indigenous territories. In New Caledonia, as seen in Chap. 7, the representation of type of water is very different from the western vision. The five Kanak waters (clan water, customary water, ancient water, pure water and collective water) are very different from the categories used in the PEP (groundwater, surface water, public water, water on customary lands). Simms et al., 2016 also insists on the role of data and sharing data. But the data used is based on a western way of quantifying or qualifying water. As seen on Chaps. 8 and 9, such a framework places significant emphasis on resource measurements, although Kanak people can also measure quality through "the health of the communities and of their links".

Another advantage, partly as a result of the Matignon-Oudinot and then Nouméa Agreements, is the fairly advanced integration of Kanak people into the governance systems, some of whom are politicians and sit in various assemblies. As in Bolivia, there is an overlap between the customary and non-customary political spheres. As a result, the Kanak use political channels to defend their interests. However, their position on water governance remains largely invisible. For a very long time, these issues, often seen as technical and tied to major agricultural economic interests, were primarily the domain of the bush loyalists (anti-independentists), who are close to the European rural world. Administratively and technically, however, some Kanak water technicians are facilitating connections between the provincial and customary authorities. The challenge lies in balancing their political commitment with their respect for customary systems. Customary representation could be strengthened to better integrate their cultural interests and demands into local policies. On the ground, agreements and tinkering between resource management methods are commonplace in the day-to-day management of drinking water networks (Peytavi et al., 2023).

In New Caledonia, while the institutional framework for governance aims to formalise relations between indigenous and non-indigenous actors, particularly in water management, the institutional solutions proposed focus on the legal aspect without distinguishing between land tenure systems. Formalising relations between indigenous and non-indigenous actors can create tensions, particularly over the unique status of water, which is separate from land tenure.

However, as early as 2011, Massenavette clarified the definition of New Caledonia's public river domain, taking into account the 2009 revision of Organic Law 99-209 (Massenavette, 2011). In 2018, David presented his thesis on the participatory construction of environmental law in the Loyalty Islands Province with customary and non-customary stakeholders (David, 2018). In June 2020, a report on water law and legal pluralism shed light on these concepts by examining cases from Bolivia, Australia and New Zealand (David & Lecercle, 2020). This provides New Caledonian policy-makers and practitioners with examples of how to incorporate indigenous values and representations into an institutional framework. Other examples from New Caledonia illustrate how indigenous representations can be taken into account. This is the case of the regulations governing the green sea turtle (Chelonia mydas) which cannot be hunted because the species is fully protected. However, given the value of this species in Kanak custom, the customary authorities can request an exemption to hunt this turtle as part of a customary ceremony. The permit specifies the place and date of the hunt or the number of turtles required. It is then up to the provincial assembly to consider the request and, if the state of the species permits, to grant the customary authority the exemption. This example illustrates how customary and non-customary practices can be formalised while respecting the values of each party (Dégremont & Sabinot, 2020). There are many examples of pluralism, but it is important to consider the construction of 'pragmatic' law, where written rules can be adapted to the resources available (Mayaux, 2019) or where the framework leaves room for informal arrangements at the local level (Riaux, 2011). However, at this stage of the process, the water law that was proposed is not on the path to inclusion. As seen in Chap. 5, the new water law is, at this stage, a missed opportunity to achieve cultural and legal pluralism.

One way to continue making practical progress toward this pluralist goal is undoubtedly to take an interest in and discuss the customary initiatives on governance and pluralism outlined in the Charter of the Kanak People and their vision of the institutional future of "Kanaky-New Caledonia". These proposals from the Kanak world have a format that is relatively compatible with the format conventionally used by administrations (written charter, articles, signature by recognised authorities). Their content attempts to reflect the specific features of Kanak social organisation, as well as Kanak values and perspectives on resources and approaches to decision-making.

The Kanak approach to producing these documents, which conforms to technocratic and bureaucratic practices, raises the question of whether this is a case of the incorporation of Kanak values and systems into Western culture, or whether it is more a case of inclusion, where the effects between the systems are two-way. The Kanak approach to the construction of the environmental code was voluntary and can be compared to inclusion, where each party influenced the other (taking into account values, adapting methods, following the protocol) (Jackson et al., 2012).

Consequently, New Caledonia has many assets with which to make progress towards more inclusive water governance, but it is really the alignment of value and organisational systems and the participation of the Kanak people in the governance system, accompanied by a coherent institutional framework, that will guarantee successful inclusion. The example of the communal mining committee provided for in Article 7 of the Mining Code, the only one of its kind in New Caledonia, demonstrates the concept of a commission comprising a diverse mix of members. This committee has a consultative role, but it also has regulatory powers over the projects carried out, particularly when several members of the committee support the same point of view. This example illustrates the importance of the power dynamics that unfold through such mechanisms, and highlights the importance of including representatives from the customary world, either appointed or elected according to customary rules. These committees are all examples of the practical implementation of a real integration of the Kanak people into the management mechanisms and demonstrate the possibility of engaging in similar dynamics for sharing different types of knowledge and data.

New Caledonia could also draw inspiration from initiatives in Canada and Australia (Maclean, 2015), such as action research programs that bring together institutional actors, indigenous stakeholders and researchers from the humanities and social sciences. These programs encourage the mutual exchange of knowledge, methods and understanding of values and practices that are essential for adapting operational governance systems to customary water use (Woodward et al., 2012). Arsenault describes the concept of the 'Indigenous Community-Based Research Lab', which aims to conduct research with indigenous people and along indigenous lines in order to align more closely with their representations of water or water organisation and management systems. The aim is to translate traditional knowledge and indigenous practices into a format that can be understood and shared, so that they can be incorporated into management plans that respond to the issues affecting indigenous peoples (Arsenault et al., 2018). In New Caledonia, this approach involving stakeholders from different levels of government, customary authorities and researchers is still in its infancy and would certainly be conducive to aligning PEP objectives with indigenous priorities on the ground. In this regard, the mutual inclusion of systems in governance structures tailored to local issues and specificities could be explored. The task of making the systems coherent and reconciling their values is a colossal one, but New Caledonia has the necessary resources and actors willing to facilitate it.

14.5 Where Do We Stand and How Can We Get Back on Track with Inclusive Governance?

New Caledonia has several assets to succeed in creating inclusive governance despite its colonial legacy. The whole challenge is the integration of indigenous peoples as an entity in its own right, and the integration of the cultural norms, values and practices of these peoples into the mechanisms and public policies as a result. Tackling this challenge requires reflection on the potential imbalances within each committee and at each level that seeks to support local collaboration between indigenous and non-indigenous stakeholders. Addressing these disparities is crucial to fostering effective governance. The idea is to progressively align regulatory frameworks with operational practices and correcting imbalances, and avoiding the pitfalls of poor and superficial incorporation.

Despite all these strengths and progress made since the 2010s, as well as the economic difficulties faced by the region following the COVID crisis and the fall in nickel prices, the water law proposed for 2023 ultimately proved to be less ambitious and inclusive, as customary lands were excluded from the text (see Chap. 5). The efforts made in the participatory approach (see Chap. 9) and the development of research to promote the integration of indigenous representations, values and ways of working were not enough.

However, the results presented in this chapter encourage further reflection and experimentation with governance mechanisms that take into account the Kanak people, their customs, knowledge and practices. Although there are many issues at stake in taking indigenousness into account in water governance, in the case of New Caledonia and its complex multi-level governance, they ultimately boil down to the distribution of competences provided for by law in a context where the government's financial resources are limited; the question of competences and the availability of people to act as a link between the institutions and the local level; and, finally, data and information asymmetries between the parties.

Nevertheless, a comparison of the New Caledonian case with situations in other countries shows that the consideration of the governance procedures and the adaptation between indigenous and non-indigenous norms and cultural values is a decisive factor in integrating the challenges of indigenousness. It is a question of proposing governance mechanisms capable of providing institutional and operational frameworks that enable peoples to participate in decision-making as entities in their own right and as bearers of their own values, representations and practices, and also to share knowledge about norms, values and practices with decision-makers. This is key to the reciprocal inclusion of different systems in an operational and institutional framework. Working at the local level, bringing together indigenous and non-indigenous stakeholders, is vital for raising awareness of each side's concerns and for reaching agreements on compatible ways of working and decision-making. Once this grassroots work has been carried out, discussions on the design of regulatory, financial and policy monitoring frameworks will move towards greater consistency between stakeholder demands, framework documents and operational implementation. The approach has yet to be tested, but it could help avoid the pitfall

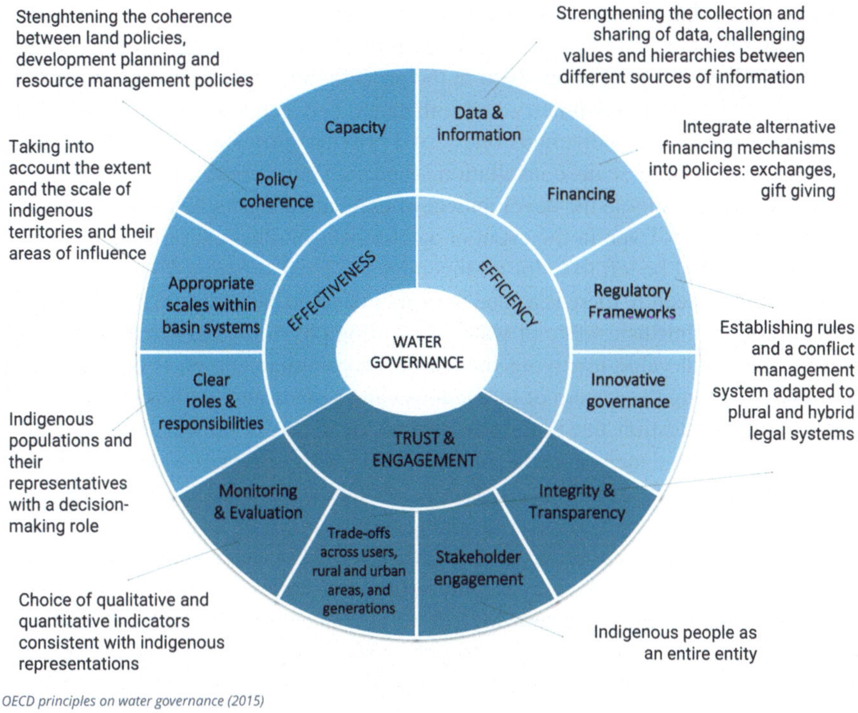

Fig. 14.2 Avenues for reinterpreting OECD principles in New Caledonia

of incorporating one system into another. New Caledonia is well placed to test this approach and continue the current momentum in water policy.

Several avenues for improvement can already be formulated to adapt New Caledonia's water governance model and make it more inclusive (Fig. 14.2). In terms of data, it is a matter of continuing the processes of creating and collecting data on water in all its diversity, regardless of the underlying ontologies, and challenging the hierarchies between types of data in guiding public decision-making. New perspectives and innovations are also likely to be found in the ways in which resource management is financed, with particular reference to non-market practices (fees, taxes or contributions in kind and labour), etc.

14.6 Conclusion and Perspectives to Improve Water Governance in Multicultural Context

The case of New Caledonia and the whole experience of the PEP reported in this book are inspiring for policy makers, especially in countries where endogenous and exogenous populations and/or legal orders coexist. It helps bring together key ideas and pathways for improving water governance.

First, in any large-scale participatory process there is a need for a debate and a consensus on the legitimacy of who is participating and talking, with a specific focus on Indigenous people involved in the process. Secondly, it is important to be aware of the potential asymmetry and imbalance between knowledges and epistemologies (Western vs. Indigenous ones), and provide sufficient financial resources to implement a large scale consultation, especially in remote areas. Finally, the approach should promote the development of innovative legal solutions and encourage legal creativity. Even in the event of a consensus on this matter, some room for manoeuvre should be left to tailor arrangements to local structures.

In terms of research, there is a need to further develop basic research and research activities on the intrinsic value of water in meeting physiological needs and its significant relational value involving human and non-human actors. It is essential to continue clarifying the system of indigenous values associated with water (Chaps. 6 and 9) and to question the dominant, technocratic model of water management based on control and measurement. In this context, the concept of water stewardship with creation of cross-fertilisation forums to facilitate shared learning between stakeholders and the construction of a language to decolonise water management advanced by Margreet Zwarteveen et al. (2024), could be placed at the heart of discussions. To effectively take into account values and representations of water, there is a need to "decolonize" western logical water frameworks, based for instance on specific categories of water (groundwater, surface water vs ancient water, pure water), on western values (such as the weight given to indicators or data to inform water quantity and quality versus the "health of the communities" as a way to inform water quality). As written by Zwarteveen et al. (2024), "*comparisons across heterogeneous communities sometimes require difficult translations and simplifications. To avoid getting trapped in one single language, we suggest nurturing and thinking with differences, learning from each other's idioms so that no one remains the same as they were at the beginning*".

Effective water management must take into account these multiple values and relational dynamics in negotiations and help to build new ways of managing different types of water, through management protocols and/or through others approaches such as "caring for water". This requires openness to the evolving assemblages around water that reflect different ideas, norms, technologies and infrastructures. By providing insights into these dynamics, this type of research would assist stakeholders in designing water policies that recognise and integrate the diverse approaches to understanding and managing water in New Caledonia.

References

Arsenault, R., Diver, S., McGregor, D., Witham, A., & Bourassa, C. (2018). Shifting the framework of Canadian water governance through indigenous research methods: Acknowledging the past with an eye on the future. *Water, 10*(1), 49. https://doi.org/10.3390/w10010049

Boelens, R., Bustamante, R., & Vos, H. de. (2007). Legal pluralism and the politics of inclusion: Recognition and contestation of local water rights in the Andes. In B. van Koppen, M. Giordano, & J. Butterworth (eds.), Community-based water law and water resource management reform in developing countries (p. 96–113). CABI. https://doi.org/10.1079/9781845933265.0096.

Bouard, S., Lejars, C., Sabinot, C., Coulange, D., Grochain, S., et al. (2020). Goutte (volet 2): La gouvernance de l'eau sur terres coutumières en Nouvelle-Calédonie: De la connaissance des usages et des savoirs à la production d'outils de gestion et de politiques adaptés. In F. Cayrol & C. Tatin (Eds.), *Au fil de l'eau, Nouméa 2019, workshop des 17 et 18 septembre* (pp. 30–33). PUNC. Collection CRESICA. Retrieved from https://hal.science/hal-02912017

Bradford, L., Ovsenek, N., & Bharadwaj, L. A. (2016). Indigenizing water governance in Canada. In *Water policy and governance in Canada* (Vol. 17, p. 269). Springer International Publishing. Retrieved from https://www.researchgate.net/publication/310114475_Indigenizing_Water_Governance_in_Canada

Canessa, A. (2014). Les paradoxes des politiques multiculturelles en Bolivie: Entre inclusion et exclusion. *Problemes d'Amerique Latine, 92*(1), 11–26. Retrieved from https://www.cairn.info/revue-problemes-d-amerique-latine-2014-1-page-11.htm

Cardenas, Y. V. (2007). La construction sociale du statut juridique de l'eau en Amérique du Nord. In D. R. Thévenot (Ed.), *Journées Scientifiques de l'Environnement* (Vol. 3, p. 13) Retrieved from https://hal.archives-ouvertes.fr/hal-00195937

Cornut, E., & Deumier, P. (2016). *L'intégration de la coutume dans le corpus normatif contemporain en Nouvelle-Calédonie* (p. 558). Université de la Nouvelle-Calédonie et Université Jean Moulin Lyon 3.

Coulange, D., Bouard, S., Grochain, S., Lejars, C., & Sabinot, C. (2021). Analyser et comprendre les représentations, les savoirs et les pratiques autour de l'eau sur terres coutumières pour appuyer la mise en œuvre de la Politique de l'Eau Partagée (PEP). *RJPENC, 1*(37), 207–221. Retrieved from https://rjpenc.nc/fr/

Cranney, K., & Tan, P.-L. (2011). Old knowledge in freshwater: Why traditional ecological knowledge is essential for determining environmental flows in water plans. *Australasian Journal of Natural Resources Law and Policy, 14*(2), 71–114. Retrieved from https://www.researchgate.net/publication/287676720_Old_knowledge_in_freshwater_Why_traditional_ecological_knowledge_is_essential_for_determining_environmental_flows_in_water_plans

David, V. (2018). *Pour une meilleure protection juridique de l'environnement en Nouvelle-Calédonie Innover par la construction participative du droit.* Thesis, Paris [Sciences et Lettres]. Retrieved from http://www.theses.fr/2018PSLEH001

David, V., & Lecercle, H. (2020). Droit de l'eau et pluralismes: contribution à la construction du droit de l'eau en Nouvelle-Calédonie, Rapport projet GOUTTE Juin 2020, https://cloud.cresica.nc/index.php/s/Kx2ToWGGJrEXmEj

Dégremont, M., & Sabinot, C. (2020). La terre-mer, enjeu de territoires et de légitimités en Nouvelle-Calédonie. *L'Information Geographique, 84*(1), 79–111. Retrieved from https://www.cairn.info/revue-l-information-geographique-2020-1-page-79.htm

Drobenko, B. (2013). L'essentiel du droit de l'eau. *Gualino*. Retrieved from https://www.unitheque.com/essentiel-droit-eau.-edition/les-carres/gualino/Livre/66500

Durette, M. (2010a). A comparative approach to Indigenous legal rights to freshwater: Key lessons for Australia from the United States, Canada and New Zealand. *Thomson Reuters, 27*, 296–315.

Durette, M. (2010b). An integrative model for cultural flows: Using values in fisheries to determine water allocations. *Synexe*. Retrieved from https://static1.squarespace.com/static/53b85a6ce4b046129f33a042/t/54715824e4b0222913423da8/1416714276844/An+integrative+model+for+cultural+flows+-+Using+values+in+fisheries+to+determine+water+allocations.pdf

Escott, H., Beavis, S., & Reeves, A. (2015). Incentives and constraints to Indigenous engagement in water management. *Land Use Policy, 49*, 382–393. https://doi.org/10.1016/j.landusepol.2015.08.003

Finn, M., & Jackson, S. (2011). Protecting indigenous values in water management: A challenge to conventional environmental flow assessments. *Ecosystems, 14*(8), 1232–1248. https://doi.org/10.1007/s10021-011-9476-0

Hassenteufel, P. (2011). *Sociologie politique: L'action publique*. Armand Colin. https://doi.org/10.3917/arco.hasse.2011.01

Hill, R., Grant, C., George, M., Robinson, C., Jackson, S., & Abel, N. (2012). A typology of indigenous engagement in Australian environmental management: Implications for knowledge integration and social-ecological system sustainability. *Ecology and Society, 17*(1), 23. https://doi.org/10.5751/ES-04587-170123

Holley, C., & Sinclair, D. (2014). *A new water policy option for Australia? Collaborative water governance, compliance and enforcement and audited self-management* (SSRN Scholarly Paper ID 3161059). Social Science Research Network. Retrieved from https://papers.ssrn.com/abstract=3161059

Hsiao, E. C. (2012). Whanganui River Agreement. *Environmental Policy and Law, 42*(6), 371–375.

Hughey, K., Jacobson, C., & Smith, E. (2017). A framework for comparing collaborative management of Australian and New Zealand water resources. *Ecology and Society, 22*(4), 18. https://doi.org/10.5751/ES-09582-220428

Jackson, S. (2018). Water and Indigenous rights: Mechanisms and pathways of recognition, representation, and redistribution. *WIREs Water, 5*(6), 1–26. https://doi.org/10.1002/wat2.1314

Jackson, S., & Altman, J. (2009). Perspectives from tropical northern Australia. *Australian Indigenous Law Review, 13*(1), 27–48. Retrieved from https://www.jstor.org/stable/26423115

Jackson, S., Tan, P.-L., Mooney, C., Hoverman, S., & White, I. (2012). Principles and guidelines for good practice in Indigenous engagement in water planning. *Journal of Hydrology, 474*, 57–65. https://doi.org/10.1016/j.jhydrol.2011.12.015

Jackson, M., Stewart, R. A., & Beal, C. (2019). Identifying and overcoming barriers to collaborative sustainable water governance in Remote Australian Indigenous Communities. *Water, 11*(2410), 33. https://doi.org/10.3390/w11112410

Jiménez, A., Saikia, P., Giné, R., Avello, P., Leten, J., Liss Lymer, B., Schneider, K., & Ward, R. (2020). Unpacking water governance: A framework for practitioners. *Water, 12*(3), 827. https://doi.org/10.3390/w12030827

Lacroix, L. (2013). Un multiculturalisme sans minorités? Quelques réflexions sur l'Etat plurinational en Bolivie et en Equateur. *Belgeo, 3*, Article 3. https://doi.org/10.4000/belgeo.11512

Lascoumes, P., & Le Galès, P. (2012). *Sociologie de l'action publique* (2nd ed.). Armand Colin. Retrieved from http://journals.openedition.org/lectures/8588

Maclean, K. (2015). Crossing cultural boundaries: Integrating Indigenous water knowledge into water governance through co-research in the Queensland Wet Tropics, Australia. *Geoforum, 59*, 142–152. https://doi.org/10.1016/j.geoforum.2014.12.008

Massenavette, C. (2011). La reconnaissance législative du domaine fluvial de la Nouvelle-Calédonie: La fin d'une longue période d'insécurité juridique. *Revue Juridique de l'Environnement, 36*(1), 17–30. Retrieved from https://www.cairn.info/revue-revue-juridique-de-l-environnement-2011-1-page-17.htm

Mayaux, P.-L. (2019). A la recherche de la réutilisation des eaux urbaines en agriculture: Rationalité technicienne et impensés institutionnels dans le projet d'irrigation de Settat-Sidi El Aidi. *Alternatives Rurales, 7*, 19.

OECD. (2015a). *Stakeholder engagement for inclusive water governance*. OECD Publishing.

OECD. (2015b). *The governance of Water Regulators*. OECD Publishing.

Peytavi, O., Bouard, S., Le Meur, P.-Y., & Lejars, C. (2023). Sociotechnical tinkering with freshwater supply: The co-creation of water knowledge in New Caledonia. *Grassroots, 30*(1), 413–423. https://doi.org/10.2458/jpe.5289

Pittet, C. (2019). *Rapport de mission sur la délégation de la gestion de la ressource en eau et la gestion du domaine public fluvial en Nouvelle-Calédonie. [Rapport de mission]* (p. 54). Gouvernement de la Nouvelle-Calédonie, DAVAR.

Poelina, A., Taylor, K. S., & Perdrisat, I. (2019). Martuwarra Fitzroy River Council: An Indigenous cultural approach to collaborative water governance. *Australasian Journal of Environmental Management, 26*(3), 236–254. https://doi.org/10.1080/14486563.2019.1651226

Riaux, J. (2011). Faut-il formaliser les règles de gestion de l'eau? Une expérience dans le Haut Atlas. *Agriculture, 20*(1–2), 67–72.

Simms, R., Harris, L., Joe, N., & Bakker, K. (2016). Navigating the tensions in collaborative watershed governance: Water governance and Indigenous communities in British Columbia, Canada. *Geoforum, 73*, 6–16. https://doi.org/10.1016/j.geoforum.2016.04.005

Tarte, F.-A. D. (2015). *L'implication des Premières Nations du Québec dans la Gestion Intégrée de l'Eau par bassin versant [Essai pour l'obtention du grade de maître en environnement]* (p. 139). Université de Sherbrooke. Retrieved from https://www.usherbrooke.ca/environnement/fileadmin/sites/environnement/documents/Essais_2015/Tarte_Felix_Antoine_D_MENV_2015.pdf

Taylor, K., Longboat, S., & Grafton, R. (2019). Whose rules? A water justice critique of the OECD's 12 principles on water governance. *Water, 11*, 809.

Tockman, J., Cameron, J., & Plata, W. (2015). New institutions of indigenous self-governance in Bolivia: Between autonomy and self-discipline. *Latin American and Caribbean Ethnic Studies, 10*(1), 37–59. https://doi.org/10.1080/17442222.2015.1034442

von der Porten, S., & de Loë, R. (2013a). Water governance and Indigenous governance: Towards a synthesis. *Indigenous Policy Journal, 23*(4), 12. Retrieved from http://blog.indigenouspolicy.org/index.php/ipj/article/view/148

von der Porten, S., & de Loë, R. C. (2013b). Collaborative approaches to governance for water and Indigenous peoples: A case study from British Columbia, Canada. *Geoforum, 50*, 149–160. https://doi.org/10.1016/j.geoforum.2013.09.001

von der Porten, S., & de Loë, R. C. (2014a). Water policy reform and Indigenous governance. *Water Policy, 16*(2), 222–243. https://doi.org/10.2166/wp.2013.0

von der Porten, S., & de Loë, R. C. (2014b). Water policy reform and Indigenous governance. *Water Policy, 16*(2), 222–243. https://doi.org/10.2166/wp.2013.046

Woodward, E., Jackson, S., Finn, M., & McTaggart, P. M. (2012). Utilising Indigenous seasonal knowledge to understand aquatic resource use and inform water resource management in northern Australia. *Ecological Management & Restoration, 13*(1), 58–64. https://doi.org/10.1111/j.1442-8903.2011.00622.x

Zwarteveen, M., Domínguez-Guzmán, C., Kuper, M., Saidani, A., Kemerink-Seyoum, J., Cleaver, F., Kulkarni, H., Bossenbroek, L., Ftouhi, H., Verzijl, A., Aslekar, U., Kadiri, Z., Chitata, T., Leonardelli, I., Kulkarni, S., & Bhat, S. (2024). Caring for groundwater: How care can expand and transform groundwater governance. *International Journal of the Commons, 18*(1), 384–396. https://doi.org/10.5334/ijc.1350

Index

C. Lejars, S. Bouard (eds.), *Water Policy in New Caledonia*, Global Issues in Water Policy 32, https://doi.org/10.1007/978-3-031-88196-1

The manufacturer's authorised representative in the EU is Springer Nature Customer Service Centre GmbH, Europaplatz 3, 69115 Heidelberg, Germany. If you have any concerns regarding our products, please contact ProductSafety@springernature.com

Printed and bound by CPI Group (UK) Ltd, Croydon, CR0 4YY
01/10/2026
02225966-0010